D1810991

Rana Majed

Thermodynamic and Kinetic Study for the Corrosion of Aluminium

Rana Majed

Thermodynamic and Kinetic Study for the Corrosion of Aluminium

LAP LAMBERT Academic Publishing

Impressum / Imprint

Bibliografische Information der Deutschen Nationalbibliothek: Die Deutsche Nationalbibliothek verzeichnet diese Publikation in der Deutschen Nationalbibliografie; detaillierte bibliografische Daten sind im Internet über http://dnb.d-nb.de abrufbar.
Alle in diesem Buch genannten Marken und Produktnamen unterliegen warenzeichen-, marken- oder patentrechtlichem Schutz bzw. sind Warenzeichen oder eingetragene Warenzeichen der jeweiligen Inhaber. Die Wiedergabe von Marken, Produktnamen, Gebrauchsnamen, Handelsnamen, Warenbezeichnungen u.s.w. in diesem Werk berechtigt auch ohne besondere Kennzeichnung nicht zu der Annahme, dass solche Namen im Sinne der Warenzeichen- und Markenschutzgesetzgebung als frei zu betrachten wären und daher von jedermann benutzt werden dürften.

Bibliographic information published by the Deutsche Nationalbibliothek: The Deutsche Nationalbibliothek lists this publication in the Deutsche Nationalbibliografie; detailed bibliographic data are available in the Internet at http://dnb.d-nb.de.
Any brand names and product names mentioned in this book are subject to trademark, brand or patent protection and are trademarks or registered trademarks of their respective holders. The use of brand names, product names, common names, trade names, product descriptions etc. even without a particular marking in this works is in no way to be construed to mean that such names may be regarded as unrestricted in respect of trademark and brand protection legislation and could thus be used by anyone.

Coverbild / Cover image: www.ingimage.com

Verlag / Publisher:
LAP LAMBERT Academic Publishing
ist ein Imprint der / is a trademark of
AV Akademikerverlag GmbH & Co. KG
Heinrich-Böcking-Str. 6-8, 66121 Saarbrücken, Deutschland / Germany
Email: info@lap-publishing.com

Herstellung: siehe letzte Seite /
Printed at: see last page
ISBN: 978-3-8484-1996-8

Zugl. / Approved by: University of Technology/Material Engineering

Thermodynamic and Kinetic Study for the Corrosion of Aluminium and Its Alloys in the Basic Media

By

Rana Afif Majed

B. Sc. Baghdad University

M. Sc. Baghdad University

April 2007

1

Symbols And Abbreviation

Symbol	Definition	Unites
A	Pre – exponential factor	$Molecules.m^{-2}.s^{-1}$
b_a	Anodic Tafel slope	$V.decade^{-1}$
b_c	Cathodic Tafel slope	$V.decade^{-1}$
DMF	Dimethyl formamide	
E	Potential	V
E_a	Activation energy	$kJ.mol^{-1}$
E_{corr}	Corrosion potential	V (SCE)
E_{pass}	Passive potential	V (SCE)
E_b	Breakdown potential	V (SCE)
F	Faraday constant	$C.mol^{-1}$
ΔG	Gibbs free energy change	$kJ.mol^{-1}$
ΔH	Enthalpy change	$kJ.mol^{-1}$
i	Current density	$A.cm^{-2}$
i_{corr}	Corrosion current density	$A.cm^{-2}$
i_{pass}	Passive current density	$A.cm^{-2}$
i_o	Equilibrium exchange current density	$A.cm^{-2}$
ipy	Inch per year	
g/Ah	Gram per Amper. hour	
MPa	Mega Pascal	
mdd	Milligram per dm^2 . day	
n	Number of electrons	
P	Protection efficiency	
r	Rate of corrosion	$A.cm^{-2}$
R	Gas constant	$J.mol^{-1}.K^{-1}$
R_p	Polarization resistance	$\Omega.cm^{-2}$
SCE	Saturated calomel electrode	V
ΔS	Entropy change	$J.mol^{-1}.K^{-1}$
T	Kelvin temperature	K
α	Transfer coefficient	
α_a	Anodic transfer coefficient	
α_c	Cathodic transfer coefficient	
β	Symmetry factor	
η	Over potential	
W/mK	Wat per meter. Kelvin	

Contents

<u>Summary</u>

The subject of this work involves the investigation of the polarization behaviour of aluminium as well as of three of aluminium – base alloys in sodium hydroxide solutions over temperature range (298 – 313 K).

The major subjects of the work and the main results obtained may be presented as follows :

1- Polarization behaviour studies of pure Al and its three alloys Al-Cu-Mg (2024), Al-Mg (5083), and Al-Zn-Mg (7075). The research was performed in three concentration of NaOH solution at pH values of 13, 11, and 9. The polarization behaviour of pure Al and its alloys have been examined using a potentiostat (Corroscript) which was obtained from Tacussel (France) at a scan rate of 0.3 Volt per minute. The main results obtained were expressed in terms of the corrosion potentials (E_{corr}) and corrosion current densities (i_{corr}). Cathodic and anodic Tafel slopes and transfer coefficients were calculated and the results were interpreted according to the rate – determining step from charge transfer process to either chemical – desorption or to electrochemical desorption.

The measurement of polarization resistance (R_p) were performed which corresponds to the resistance (R) of the metal/solution interface to charge – transfer reaction. Polarization resistance for pure aluminium is greater than its values for alloys in three value of pH (13, 11, and 9), and the sequence of (R_p) for aluminium alloys were in the order as follow:

$$(R_p) \quad 7075 > 5083 > 2024$$

This means that Al-Zn-Mg alloy was more resistance to corrosion than other alloys.

Polarization resistance varies with the concentration of NaOH solution as follow:

$$(R_p) \text{ at } pH \qquad 13 > 9 > 11$$

This means that the medium at pH=11 was more corrosive than the medium at pH =13 and 9. Finally (R_p) values decreases with the increasing of temperature.

2- Study the effect of chloride ions, the corrosion potentials and corrosion current densities also were changed considerably by the presence of (1×10^{-3}, 1×10^{-2}, and 0.1 mol.dm^{-3}) of Cl$^-$ ions in the basic medium over the temperature range (298 – 313 K).

The presence of Cl$^-$ at pH=13 caused a shift in the (E_{corr}) to more negative values and the (i_{corr}) to the higher values. Generally, presence of Cl$^-$ in solution at pH=11 shift (E_{corr}) to more negative values and (i_{corr}) to lower values, while the presence of Cl$^-$ in solution at pH=9, in general, shift (E_{corr}) either to more or less negative values and (i_{corr}) to lower values.

The effect of chloride ion can be observe during the polarization resistance values which follow the following sequence at pH=13 with three concentrations of Cl$^-$:

$$R_p \qquad 7075 > 2024 > 5083 > pure \ Al$$

This means that the Al-Zn-Mg alloys gives more resistance to the presence of chloride ions in solution than the pure Al and other alloys.

While at pH=11 and 9, the presence of chloride ions in solution gives different behaviour as shown bellow :

In pH=11 (1×10^{-3} and 1×10^{-2} mol.dm^{-3} Cl$^-$)   *pure Al > 7075 > 5083 > 2024*

(0.1 mol.dm^{-3} Cl$^-$)   *7075 > 5083 > 2024 > pure Al*

8

In pH=9   (1x10^{-3} mol.dm^{-3} Cl$^-$)   *pure Al > 2024 > 5083 > 7075*

                 (1x10^{-2} and 0.1 mol.dm^{-3} Cl$^-$)   *pure Al > 5083 > 2024 > 7075*

3- The corrosion protection was investigated for pure Al and its three alloys in the basic medium of a temperature range (298 – 313 K) using two additives which involved sodium acetate (as organic inhibitor) in the concentration range (0.05 – 0.15 mol.dm^{-3}) and sodium chromate (as inorganic inhibitor) in the concentration range (5x10^{-3} – 5x10^{-2} mol.dm^{-3}).

Values of (P%) were negative for pure Al and its alloys in 0.1 mol.dm^{-3} NaOH solution (pH=13) in presence of CH$_3$COONa and Na$_2$CrO$_4$ with three experimental concentration because of high concentration of medium.

The presence of sodium acetate in pH=9 gives good protection efficiency (P%) for pure Al and its alloys and higher than the corresponding values in pH=11. While the presence of sodium chromate in pH=9 gives good protection efficiency for pure Al and its alloys but lower than that was observed in the pH=11. The extent of inhibition decreased with a rise in temperature, except for some certain cases.

4- Values of the thermodynamic quantities (ΔG, ΔS, and ΔH) were calculated for pure Al and its alloys in the absence and presence of additives in basic media.

Generally, values of ΔG are negatively suggesting the existence of thermodynamic feasibility for the corrosion of the electrodes materials in NaOH solution in the absence or the presence of additives. Values of ΔS are generally positive, this suggests a lower order in the solvated state of metal ions as compared with the state of metal atoms in crystal lattice of corroding electrodes.

ΔH ranges from negative to positive values indicating exothermic or endothermic nature of corrosion reaction.

5- The kinetic data have been obtained from the corresponding corrosion current densities were controlled by Arrhenius type rate equation.

Values of E_a and log A varies with variation of electrodes (metal and alloys) and pH values.

The results of a linear relationship existed between the values of E_a and log A were identical to the compensation effect that the simultaneous increases or decreases E_a and log A for a system tend to compensate from the stand point of the reaction rate. The effect ascribed to the presence of energetically heterogeneous reaction sites on the alloy surface suffered corrosion in the electrolytic solution.

Simultaneous increase or decrease in E_a and log A implies a higher or lower rate. A compensation which operates is possible for striking variations in E_a and log A sites on a metal or an alloy to yield only a small variation in reactivity with a different pH and additives.

Chapter One: Theoretical Part

1-1 General Introduction

Corrosion is the interaction of a material with its environment [1-3], it is also considered as any process that involves the transfer of atoms (metallic) to ionic state. Corrosion involves the destructive attack of metal by chemical or electrochemical reaction with its environment[4]. Usually, the corrosion process consists of a set of redox reactions which are electrochemical in nature. Thus, the metal is oxidized to corrosion products at anodic sites and some species are reduced at cathodic sites.

Thermodynamic considerations determine whether or not a reaction can occur[2]. However, in spite of this limitation, thermodynamic is very important to an understanding of the electrochemistry of corrosion.

The kinetics of electrochemical reaction is based largely on the mixed potential theory of electrode kinetics as started by Wagner and Traud[5]. The basic assumptions of the theory are quite simple: *(a)* The kinetics of the various partial reactions can be treated separately and *(b)* no net current flows from an electrode which is in equilibrium or at steady state. The condition of no net current flow means that the total rate of reduction must equal the total rate of oxidation on the electrode surface.

When a reaction is forced a way from equilibrium, when one direction of the reaction is favored over the other, the potential at which the reaction is occurring changes.

The amount by which the potential changes is the over voltage which is defined as:

$$\eta = E - E_{eq} \qquad\qquad \ldots\ldots\ldots(1\text{-}1)$$

11

η= over voltage, E_{eq}= equilibrium potential, E= polarized potential which corresponds to corrosion potential of the metal.

The current applied to cause the departure from equilibrium is the net rate of reaction, thus,

$$i_{app} = \sum \vec{i} - \sum \overleftarrow{i} \qquad \ldots\ldots\ldots(1\text{-}2)$$

where $\vec{i}$ is the anodic current density, $\overleftarrow{i}$ is the cathodic current density and i_{app} is the applied current density.

The rate of the anodic reaction at a potential E is given by [6]:

$$\vec{i} = i_o \exp\left[\frac{E - E_{eq}}{b_a}\right] \qquad \ldots..(1\text{-}3)$$

where b_a is the Tafel slope and i_o is the equilibrium exchange current density. Similarly, for the cathodic reaction

$$\overleftarrow{i} = i_o \exp\left[\frac{E_{eq} - E}{b_c}\right] \qquad \ldots\ldots.(1\text{-}4)$$

At the corrosion potential E_{corr} the net current i becomes zero, since $\vec{i} = \overleftarrow{i}$. Thus, corrosion current density i_{corr}, is defined by:

$$i_{corr.} = i_o \exp\left[\frac{E_{corr.} - E_{eq}}{b_a}\right] - i_o \exp\left[\frac{E_{eq} - E_{corr.}}{b_c}\right] \qquad \ldots..(1\text{-}5)$$

The Tafel equation for cathodic process can be expressed in the form [7]:

$$\eta_c = \frac{RT}{\alpha_c ZF} \ln i_o - \frac{RT}{\alpha_c ZF} \ln i_c \qquad \ldots\ldots..(1\text{-}6)$$

Similarly, the activation overpotential of an anodic process is given by :

$$\eta_a = \frac{RT}{\alpha_a ZF} \ln i_o - \frac{RT}{\alpha_a ZF} \ln i_a \qquad \text{.......(1-7)}$$

where (F) is Faraday constant,(Z) is number of electrons and α is the transfer coefficient

when $\eta=0$, $i=i_{corr}$ and hence i_{corr} can be obtained by extrapolating the Tafel line to $E=E_{corr}$.

The corrosion process with the two coupled electrochemical reactions under activation polarization can be put[8,9] as :

$$i = i_{corr}\left[\exp\left(\frac{\alpha_a F\eta}{RT}\right) - \exp\left(\frac{\alpha_c F\eta}{RT}\right)\right] \qquad \text{.....(1-8)}$$

The basic[10] law of charge – transfer reactions has been expressed through the (Butler – Volmer) electrodic equation as :

$$i = i_o\left[\exp\left(\frac{\overleftarrow{\alpha} F\eta}{RT}\right) - \exp\left(\frac{\overrightarrow{\alpha} F\eta}{RT}\right)\right] \qquad \text{..........(1-9)}$$

This generalized equation (eq. 1-8) is seen to be of the same form as the simple (Butler – Volmer) equation (eq. 1-9) as:

$$i = i_o\left[\exp\left(\frac{(1-\beta)F\eta}{RT}\right) - \exp\left(\frac{-\beta F\eta}{RT}\right)\right] \qquad \text{.......(1-10)}$$

Where (i) is the measured current density , i_o is the equilibrium exchange current density, β the symmetry factor related to the potential drop through the electrochemical double layer.

At high – field approximation, of the (Butler – Volmer) equation contains two terms, one representing the de – electronation – current density $\overleftarrow{i}$ and the other, the electronation – current density $\overrightarrow{i}$, where $\overleftarrow{i} = i_o e^{(1-\beta)F\eta/RT}$ and $\overrightarrow{i} = i_o e^{-\beta F\eta/RT}$.

13

In a corrosion process the electronation – current density $\overleftarrow{i}$ decreases while the de – electronation – current density $\overrightarrow{i}$ increases.

When η is large enough $\overrightarrow{i} \gg \overleftarrow{i}$ and the $\overleftarrow{i}$ became is so small that can be dropped out of the expression. Thus, the high – field approximation of the Butler – Volmer equation (Valid at η's greater than about 0.1V) yield

$$i = i_o e^{(1-\beta)F\eta/RT} \qquad \ldots\ldots(1\text{-}11)$$

When the overpotential η is 0.01V or less for one electron transfer reaction the Butler – Volmer equation may be reduced to the form:

$$i = \frac{i_o F\eta}{RT} \qquad \ldots\ldots(1\text{-}12)$$

Equation (1-12) provides a simple way of understanding non- polarizable and polarizable interfaces.

A modified form of this equation is:

$$\frac{\eta}{i} = \frac{RT}{i_o F} = R_p \qquad \ldots\ldots(1\text{-}13)$$

So, the term $\frac{\eta}{i}$ corresponds to the polarization resistance R_p of the interface at an electrode to the charge transfer reaction.

The theoretical basis for electrochemical corrosion testing is derived from the mixed potential theory[5], described for the first time by Wagner and Traud and then developed by Stern and Co – workers[11, 12] exhibits a number of limitations. The various fundamental and experimental factors which contributed to in accuracy in the polarization resistance technique have been reviewed by Callow et. al[13].

The various papers have been published discussing new ways of interpretation of polarization data near the corrosion potential. The two – and three methods have been developed Barnartt[14, 15] for determination of the kinetics parameters for mono electrodes as well as for according electrodes. Reeve and Bech – Nielsen[16] have discussed the theoretical bases of Tafel slope determination from measurements in the range of linear polarization.

A graphical method of linear polarization curve analysis has been proposed by Oldham and Mansfeld[17]. The methods permit determination of i_{corr} without the knowledge of Tafel slopes. Periassamy and Knshnaswamy[18] have presented a graphical – numerical method for the determination of Tafel slopes and the corrosion current density from the polarization data near the corrosion potential, an advantage of the method consists in the utilization of all experimental data for the determination of mean values of i_{corr} and Tafel slopes, whereas the disadvantages lie in high time consumption and tedious calculations.

Among the methods discussed, that described by Barnartt seems to be the most attractive, due to the simplicity of the required calculations.

In a later paper, an alternative four – point method is described[19], being a modification of the Barnatt method. It provide simple and more accurate and precise calculations of the corrosion current density and Tafel slopes than the use of the Stern – Geary equation or of the three – point Barnartt method.

A new computational method, for analyzing electrode kinetic data is described. The method is based on numerical integration and differentiation of the experimental data to formulate three independent relationships in order to solve for corrosion rate (i_{corr}) and Tafel slopes[20].

1-2 Introduction To Aluminium And Aluminium Alloys

The unique combinations of properties provided by aluminium and its alloys make aluminium one of the most versatile, economical, and attractive metallic materials for a broad range of uses – from soft, highly ductile

15

wrapping foil to the most demanding engineering applications. Aluminium has a density of only (2.7 g/cm^3).

Aluminium resists the kind of progressive oxidization that causes steel to rust away. The exposed surface of aluminium combines with oxygen to form an inert aluminium oxide film only a few ten – millionths of an inch thick, which blocks further oxidation. And, unlike iron rust, the aluminium oxide film does not flake off to expose a fresh surface to further oxidation. If the protective layer of aluminium is scratched, it will instantly reseal itself.

The thin oxide layer itself clings tightly to the metal and is colorless and transparent – invisible to the naked eye. The discoloration and flaking of iron and steel rust do not occur on aluminium. Appropriately alloyed and treated, aluminium can resist corrosion by water, salt, and other environmental factors, and by a wide range of other chemical and physical agents.

Aluminum surfaces can be highly reflective. Radiant energy, visible light, radiant heat, and electromagnetic waves are efficiently reflected, while anodized and dark anodized surfaces can be reflective or absorbent.

The reflectance of polished aluminium, over a broad range of wave lengths, leads to its selection for a variety of decorative and functional uses. Aluminum typically displays excellent electrical and thermal conductivity, but specific alloys have been developed with high degrees of electrical resistively. These alloys are useful in high – torque electric motors.

Aluminum is often selected for its electrical conductivity, which is nearly twice that of copper on an equivalent weight basis. The requirements of high conductivity and mechanical strength steel – cored reinforced transmission cable.

The thermal conductivity of aluminium alloys, about 50 to 60% that of copper, is advantageous in heat exchangers, evaporators, electrically heated appliances and utensils, and automotive cylinder heads and radiator.

Aluminium is nonferromagnetic, a property of importance in the electrical and electronics industries.

It is nonpyrophoric, which is important in applications involving inflammable or explosive – materials handling or exposure. Aluminium is also non – toxic and is routinely used in containers for food and beverages. It has an attractive appearance in its natural finish, which can be soft and lustrous or bright and shiny. It can be virtually any color or texture.

The ease with which aluminium may be fabricated into any form is one of its most important assets. Often is can compete successfully with cheaper materials having a lower degree of workability. The metal can be cast by any method known to foundry men. It can be rolled to an desired thickness down to foil thinner than paper.

Aluminium sheet can be stamped, drawn, spun, or roll formed. The metal also may be standard into cable of any desired size and type.

There is almost no limit to the different profiles (shapes) in which the metal can be extruded[21].

1-3 Aluminium Alloys

The mechanical, physical, and chemical properties of aluminium alloys depend upon composition and microstructure. The addition of selected elements to pure aluminium greatly enhances its properties and usefulness. Because of this, most applications for aluminium utilize alloys having one or more elemental additions. The major alloying addition used with aluminium are copper, manganese, silicon, magnesium and zinc. The total amount of these elements can consist up to (10%) of the alloy composition (all

17

percentages given in weight percent unless otherwise noted). Impurity elements are also present, but their total percentage is usually less than (0.15%) in aluminium alloys[21].

It is convenient to divide aluminium alloys into two major categories: wrought compositions and cast compositions. A further differentiation for each category is based on the primary mechanism of property based on phase solubilities[21].

These treatments include solution heat treatment, quenching, and precipitation, or age, hardening. For either casting or wrought alloys, such alloys are described as heat treatable. A large number of other wrought compositions rely instead on work hardening through mechanical reduction, usually in combination with various annealing procedures for property development. These alloys are referred to as work hardening.

Some casting alloys are essentially not heat treatable and are used only in as – cast or in thermally modified conditions unrelated to solution or precipitation effects[21].

Their alloy identification system employs different nomenclatures for wrought and cast alloys, but divides alloys into families for simplification. For wrought alloys a four – digit system is used to produce a list of wrought composition families as follows[21] :-

1xxx :- Controlled unalloyed (pure) composition, used primarily in the electrical and chemical industries.

2xxx :- Alloys in which copper is the principal alloying element, although other elements, notably magnesium, may be specified. 2xxx – series alloys are widely used in aircraft where their high strength (yield strengths as high as 455 MPa) are valued.

3xxx :- Alloys in which manganese is the principal alloying element, used as general – purpose alloys for architectural applications and various products.

4xxx :- Alloys in which silicon is the principal alloying element, used in welding rods and brazing sheet.

5xxx :-Alloys in which magnesium is the principal alloying element, used in boat hulls, gangplanks, and other products exposed to marine environments.

6xxx :- Alloys in which magnesium and silicon are the principal alloying elements, commonly used for architectural extrusions.

7xxx :- Alloys in which zinc is the principal alloying element (although other elements, such as copper, magnesium, chromium, and zirconium, may be specified), used in aircraft structural components and other high – strength applications. The 7xxx series are the strongest aluminium alloys, with yield strengths ($\geq$ 500 MPa) possible.

8xxx :- Alloys characterizing miscellaneous compositions. The 8xxx series alloys may contain appreciable amounts of tin, lithium, and / or iron.

9xxx :- Reserved for future use[21].

1-4 Physical Properties

Some of more useful physical properties of aluminium are given in Table (1-1). The common wrought forms are rolled plate, clad plate, sheet and strip, clad sheet and strip, bars, rods, and sections, extruded round tube and hollow sections, drawn tubes, wire, rivet stock, bolt and screw struck, and forgings and forgings stock[21]. Castings are made in sand moulds or in metal moulds known as dies; the most widely used methods involve casting either under gravity or under pressure. Aluminium and aluminium alloys are fabricated into products such as rolled plate, sheet, extruded sections, drawn tube, etc. by all the familiar processes, with modifications appropriate to the temper or condition of the material. Joining may be carried out by mechanical methods (such as riveting and bolting), brazing, soldering, adhesive bonding, or welding. The argon – shielded are welding methods are particularly appropriate where corrosion resistance of welded joints is of importance[22].

Table (1-1) : *Properties of aluminium.*

Physical	
Atomic number	13
Atomic volume	10 cm^3
Atomic weight	26.97 gm
Valency	3
Crystal structure	Face – centered cubic (f.c.c)
Interatomic distance	2.863 A$^\circ$
Electrochemical equivalent	0.3354 g/Ah
Density at 293 K	2700 kg/m^3

Thermal					
m.p *(K)*	*sp. Heat at 293K (J/kg.K)*	*Mean sp. heat (293-931K) (J/kgK)*	*Latent heat of fusion (kJ/kg)*	*Coeff. Of linear exp. (293-393K) (1/K)*	*Thermal conductivity at 273K (W/mK)*
931	896	1047	387	0.61x10^{-6}	214

Electrical			
Elec. vol Resistively at 293K (µΩ cm)	*Elec. vol. conductivity at 293K (%I.A.C.S)*	*Temp. coeff. of elec. resistance per K for 293K*	*Thermoelectric power vs platinum (mV/100K)*
2.7 – 3.0	63 - 57	0.0041	+ 0.41

1-5 Corrosion Resistance Of Aluminium And Aluminium Alloys

Aluminium and most aluminium alloys have good corrosion resistance in natural atmospheres, fresh water, seawater, many soils, many chemicals and their solutions, and most foods. This resistance to corrosion is the result of the presence of a very thin, compact, and adherent film of aluminium oxide on the metal surface[21].

Whenever a fresh surface is created by cutting or abrasion and is exposed to either air or water, a new film forms rapidly, growing to a stable thickness. The film formed in air at ambient temperature is $\approx$ 5nm (50 A$^{\circ}$) thick. The thickness increases with increasing temperature and in the presence of water.

The oxide film is soluble in alkaline solutions and in strong acids, with some exceptions, but is stable over a pH range of ($\approx$ 4.0 to 9.0).

There are different types of corrosion and various interactions with induced or imposed stresses so that the effects can range from unimportant to highly damaging[21].

For some types of applications, a distinction should be made between appearance and durability. The surface can become unattractive because of roughening by shallow pitting and can darken with dirt retention, but these conditions may have no effect on durability or function. On the other hand, stress – corrosion cracking (SCC) or highly localized, severe corrosion due to heavy – metal ions in solutions, stray electrical currents, or galvanic couples with more – anodic metals can be quite damaging.

Good design and application practices must be observed to avoid these conditions. This includes selection of alloys appropriate for the conditions of the application[21].

1-6 Effects Of Alloy Composition
1-6-1 Effect Of Copper

An earlier works reported that copper reduces the pitting initiation resistance of aluminium and therefore it was advised to add small quantities of copper (up to 0.15%) to aluminium in order to reduce the pits depths by other increasing pits density[23, 24].

On the other hand Galvele and Micheli (1970)[25] and Muller and Galvele (1977)[26] studied the effect of copper on E_b value for aluminium in different concentrations of NaCl solutions and on both (1M NaCl) and saturated $AlCl_3$ solutions respectively.

They found that the presence of copper in solid solution in aluminium shifts E_b to more noble values, however this denficial effect is limited by the solubility of copper in aluminium.

Figure (1-1)[26]shows the effect of Cu – content in solid solution on E_b value for Al – Cu alloys.

They attributed the effect of copper to the preferential dissolution of aluminium from the attacked region leading to Cu – enrichment inside the pits which can reduce the cathodic reaction overpotential.

The above results were supported by Bicelli et. al. (1979)[27] who studied the effect of copper on E_b value for aluminium in 3.5% NaCl solution. They found that E_b value for Al – Cu alloy in the solution treated condition followed by natural ageing was slightly above that for commercially pure aluminium (99.9% Al).

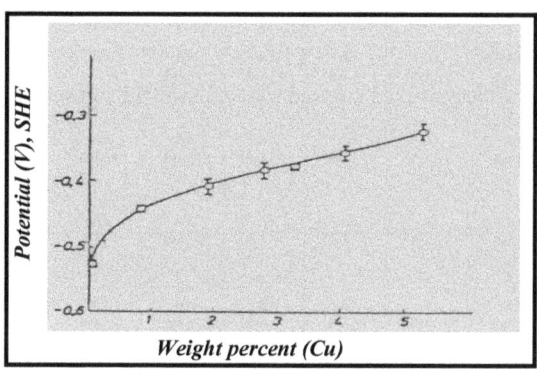

Fig. (1-1) : *Effect of the copper content on the pitting potential of Al – Cu alloys. Mean pitting potential values measured on solution treated Al – Cu alloys in de-aerated 1M NaCl solution at 25°C* [26].

1-6-2 Effect Of Zinc

Bonora et. al. (1974)[28] studied the corrosion behaviour of Al – Zn alloys in (0.5 M) NaCl solution and compared this behaviour with pure aluminium. They found that E_b decreases continuously with increasing Zn – content.

E_b values were found to be (-0.735, -0.768, -0.952, and -0.962 V(SCE)) for 99.9% Al, Al – 0.5%Zn, Al – 5%Zn, and Al – 10%Zn respectively.

Muller and Galvele (1977)[29] also studied the effect of zinc on E_b value for aluminium in (1M) NaCl and in saturated AlCl$_3$ solutions. They observed that the presence of zinc up to 3% shifts E_b to more active values but a higher increasing in Zn – content did not affect E_b value as shown in Fig. (1-2).

23

Their attribution for this effect was the preferential dissolution of aluminium from the pit region leading to Zn – enrichment inside the pit which increase the cathodic reaction overpotential causing a drop in E_{corr}.

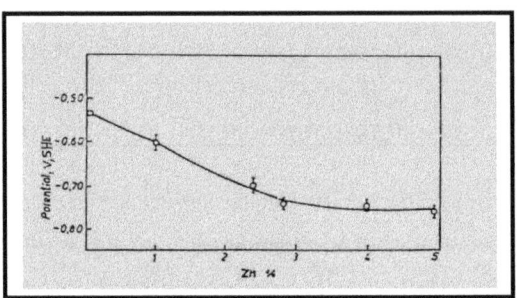

Fig. (1-2) : *Effect of the zinc content on the pitting potential of Al – Zn alloys in de- aerated (1M) NaCl solution at $25^{o}C^{(29)}$.*

1-6-3 Effect Of Magnesium

Bohni and Uhlig (1969)[30] studied the effect of magnesium on E_b value for aluminium in (0.1M) NaCl solution at 25°C. They found that the addition of (2.4% Mg) shifts E_b value from (-0.40V) to (-0.44V).

This result was confirmed by Nilsen and Bardal (1977)[31] who also studied the effect of magnesium on E_b value for aluminium in artificial sea water at 20°C. They found that magnesium shifts E_b to more active values.

Horst and English (1977)[32] studied the behaviour of Al – Mg alloy (5082) in different solutions. They observed that (Mg_2Al_3) phase has an active E_b values. On the other hand Muller and Galvele (1977)[29] studied the effect of magnesium on E_b value for different Al – Mg alloys in (1M) NaCl and saturated $AlCl_3$ solutions at 25°C.

24

By using the potentiokinetic method they found that magnesium had no effect on E_b value but the scratch method indicated that E_b value decreases continuously with increasing Mg – content as shown in Fig. (1-3). They concluded that magnesium had no effect on the pitting susceptibility of aluminium and attributed this effect to the preferential dissolution of magnesium from the surface producing an Al – rich surface leading to behaviour similar to that of pure aluminium.

This conclusion is supported by Sanad et. al. (1982)[33] who studied the effect of magnesium on the pitting resistance of aluminium in (32.7 g/L) NaCl solution.

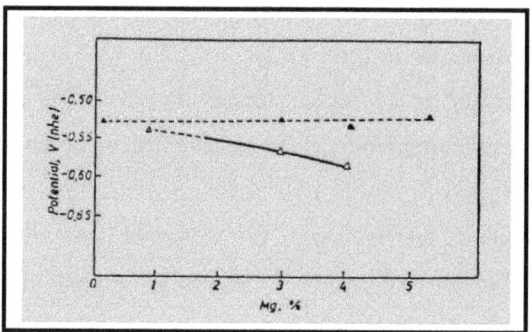

Fig. (1-3) : *Effect of the Mg content on the pitting potential of Al – Mg alloys in a de – aerated 1M NaCl solution at 25°C.*

▲ *potential measured by the potential-step technique.*
Δ *potential measured by the scratching technique* [29].

1-7 Corrosion In Atmospheres

Alloys other than those with the higher copper content have excellent resistance to atmospheric corrosion (often called weathering) and in many outdoor applications require no protection or maintenance. Products widely used under such conditions include electrical conductors.

These often retain a bright metallic appearance for many years but may darken with mild surface roughening caused by shallow pitting and with accumulation of dirt[21].

Aluminium alloys are relatively insensitive to the concentrations of oxygen present in most aqueous solutions. In general, high concentrations of dissolved oxygen tend to stimulate attack some what, especially in acid solution, although this effect is less pronounced than for most of the other common metals. As shown in Table (1-2).

Carbon dioxide or hydrogen sulfide, even in high concentrations, appears to have a slight inhibiting action on the effect of aqueous solutions on aluminium alloys.

Water solutions of hydrogen, nitrogen and chloride are strongly corrosive to aluminium. Hydrogen and nitrogen are without effect, except as they influence the oxygen content[34].

Table (1-2) _: Effect of gas atmosphere on the corrosion rate of aluminium._
Test run on three speciments[34]_._

Immersed area of speciment – 37.8 sq cm., Volume of solution -150 ml., Temperature – room., Rate of aeration -15 ml per min.

Solution	Gas	Duration of test	Corrosion rate	
			ipy	mdd
1.0% Na_2CO_3	Oxygen bubbled through solution	18 hr	0.17	320
1.0% Na_2CO_3	Air bubbled through solution	18 hr	0.17	320
1.0% Na_2CO_3	Nitrogen bubbled through solution	18 hr	0.18	338
0.1% NaOH	Oxygen bubbled through solution	6 hr	0.65	1220
0.1% NaOH	Air bubbled through solution	6 hr	0.64	1200
0.1% NaOH	Nitrogen bubbled through solution	6 hr	0.63	1190
0.1% HCl	Oxygen bubbled through solution	48 hr	0.066	124
0.1% HCl	Air bubbled through solution	48 hr	0.036	68
0.1% HCl	Nitrogen bubbled through solution	48 hr	0.006	11
0.1% HCl	Air over quiescent solution	48 hr	0.007	13

1-8 Corrosion In Water

Corrosion resistance of aluminium in high purity water, distilled or deionized, and in steam condensate is so high that these fluids are regularly contained and handled in aluminium equipment. Resistance is also high in most natural fresh waters[21].

Soft water have the least pitting tendency. Components of natural waters that increase pitting are copper ions, bicarbonate, chloride, sulfate, and oxygen. Thus, harder waters with more bicarbonate have a higher pitting tendency. Service experience with aluminium alloys in marine and coastal applications, including structures, buoys, pipelines, lifeboats, motor launches, cabin cruisers, patrol boats, barges and larger vessels, has demonstrated their

27

good resistance and long life under conditions of partial, intermittent, and total immersion.

Wrought alloys of the 3xxx, 5xxx, and 6xxx groups are used. Those alloys of the 5xxx aluminium – magnesium group are most resistance and most widely used because of favorable strength and good weldability.

The rate of corrosion based on weight loss does not exceed ≈ 5µm/year (0.2 mil/year), which is <5% of the rate for unprotected low – carbon steel in seawater[21].

Corrosion is mainly pitting decelerating with time from rates of 2.5 to 5 µm/year (0.1 to 0.2 mil/year) in the first year to average rates over a 10 year period of 0.75 to 1.5 µm/year (0.03 to 0.06 mil/year).

The curve of maximum depth of pitting versus time follows an approximate cube – root, from which it follows that doubling material thickness increases time to perforation by a factor of 8 [21].

1-9 Corrosion In Chemical Environments

Aluminium alloys are used in storing, processing, handling, and packaging of a variety of chemical products. They are compatible with most dry inorganic salts. Within the passive pH range, about 4 to 9, the resist corrosion in solutions of most inorganic chemicals but are subject to pitting in aerated solutions, particularly of halides[21].

Aluminium alloys are not suitable for containing or handling mineral acids with the exception of nitric acid in concentrations over 82 wt% and sulfuric acid in concentrations from 98 to 100 wt%. Figures (1-4) and (1-5) give corrosion data at various concentrations for these two acids[35].

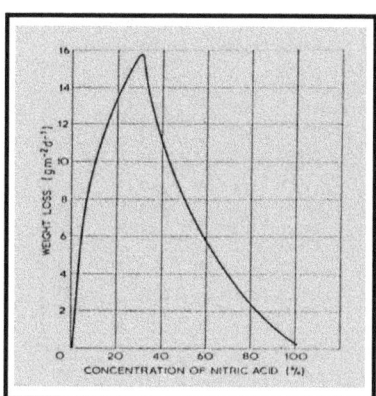

Fig. (1-4) : _Action of nitric acid_
of various conc. on commercial
purity aluminium at 20°C.

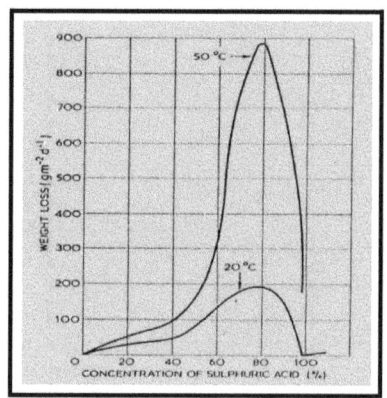

Fig. (1-5) : _Action of sulphuric acid_
of various conc. on commercial
purity aluminium at 20 and 50°C.

The corrosion rates of aluminium in other inorganic acids in dilute solution are shown in Fig. (1-6). Boric acid also exerts little attack on aluminium, while a mixture of chromic and phosphoric acids can be used for the quantitative removal of corrosion products from aluminium without attacking the metal. The effect of commercial metal purity (impurities mainly iron and silicon) on corrosion by 40% hydrochloric acid is shown in Fig. (1-7). This curve is typical of that obtained with many acids.

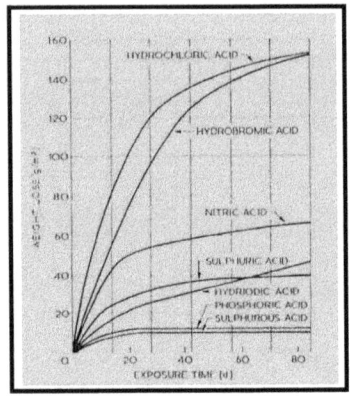

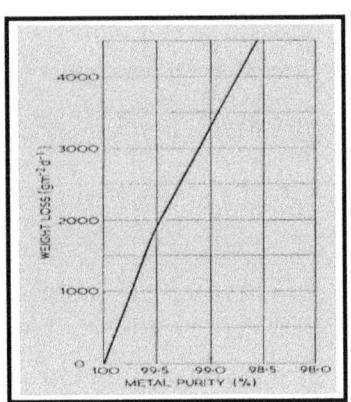

Fig.(1-6) *: Action of dilute (0.1N)*
solutions of inorganic on commercial
purity aluminium at 25°C.

Fig.(1-7) *: Action of 40% HCl*
acid on aluminium of various
purities at 25°C.

Organic acids usually have low rates of attack on aluminium, notable exceptions to this generalization being formic acid, oxalic acid and some chloride – containing acids such as trichloroacetic acid. Corrosion rates for dilute organic acid solutions are given in Fig. (1-8)[35].

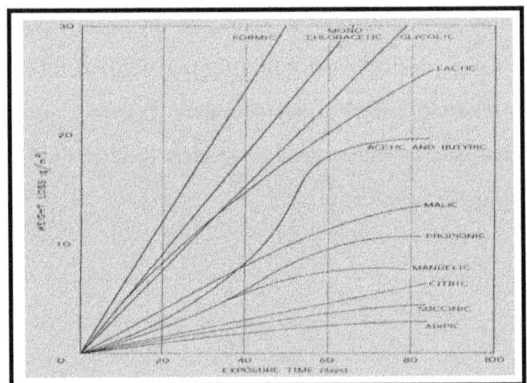

Fig. (1-8) *: Action of dilute (0.1N) solutions of*
organic acids on commercial – purity aluminium at 25°C.

Glacial acetic acid (pH=3) has no significant corrosion effect on aluminium but the rate of attack increases rapidly with decreasing concentration or in the absence of the traces of water normally present.

The rate of corrosion in an acid solution rises rapidly with temperature, often doubling or more with a 10°C rise[35].

Alkalis are generally corrosive to aluminium; caustic soda is in fact used for chemical milling of aluminium. 99% aluminium is, however, resistant to ammonium hydroxide, even at pH=13, while the action of more dilute caustic alkalis can be inhibited by the use of silicates.

Mild alkalis such as sodium carbonate are moderately corrosive and are not recommended for washing aluminium hollow – ware. Synthetic detergents, in general, give satisfactory service in cleaning aluminium, but those containing uninhibited sodium carbonate may give some surface roughening[35].

Inhibitors such as silicates can present attack by the more dilute solutions. Alloys of aluminium with magnesium or magnesium and silicon are generally more resistant than other alloys to alkaline media.

Most simple inorganic salt solutions cause virtually no attack on aluminium – base alloys, unless the possess the qualities required for pitting corrosion, which have been considered previously, or hydrolyses in solution to give acid or alkaline reactions[35].

With salts of heavy metals – notably copper, silver, and gold the heavy metal deposits on to the aluminium, where it subsequently causes serious bimetallic corrosion. Some salts, notably chromates, dichromates, silicates, borates and cinnamates, have marked inhibitive power and are very effective in closed – circuit water systems[35].

Care must be taken to ensure that a sufficient quantity of such anodic inhibitors as chromates is added, as otherwise attack, though occurring at

31

fewer points, may be more sever at these points. Chromates and dichromates have little inhibitive power in strongly acid solutions[35].

Aluminium is used in hydrogen peroxide processing and storage equipment partly because of its high corrosion resistance but also because it does not cause degradation of the peroxide.

With many organic compound, aluminium shows high corrosion resistance either in the presence or absence of water. The lower alcohols and phenols are corrosive when they are completely anhydrous – rarely encountered in practice – since repair of breaks in the natural protective oxide film on aluminium cannot take place in the absence of water. Amines generally cause little attack unless very alkaline[35].

Processing and storage equipment for many chemicals, including acetaldehyde, formaldehyde, nylon salt, methyl methacrylate, carbon tetrachloride, glycerol, triacetin, proprionic acid, acetic acid and acetic anhydride, is manufactured from aluminium alloys, primarily because of their excellent corrosion resistance.

With complete success in contact with aluminium, e.g. in aeroengines, but difficulties with graphitization of cast iron engine components in the solution have led to the introduction of two other types of inhibitors : *(a)* benzoate plus nitrite, and *(b)* borax, usually with soluble oils. Service experience has indicated that corrosion of aluminium components in these inhibited solutions occasionally takes place, though most trials give satisfactory results[35].

In refrigerating systems, halogen derivatives of methane and ethane marketed under the trade names of Arctons and Freons are without action on pure aluminium and its copper – free alloys in dry conditions, but in wet conditions monochlorodi- , dichloromono- and trichloromonofluoromethanes can hydrolyse to produce slight attack on the aluminium. Aluminium has good resistance to petroleum products, and an Al – 2Mg alloy is used for tank heating coils in crude – oil carriers. Caked – on deposits must be removed from the coils by hot sea – water cleaning in order to maintain effective heat transfer and prevent corrosion.

Aluminium is also used in the petroleum industry for sheathing for towers, heat exchangers, transport and storage tanks and scrubbers[35].

1-10 Form Of Corrosion In Aluminium

Form or type of corrosion can be categorized by the morphology, which may or may not be related to the microstructure of the metal, or by the conditions causing the corrosion. Uniform attack or dissolution, which may occur if the surface oxide film is soluble in the corroding medium, is infrequent in service[21].

Most corrosion in service is localized in one way or another. When the oxide film is insoluble in the corroding medium, corrosion is localized at weak spots in the film, which can result from microstructural features such as the presence of microconstituents.

Local cells are formed by such non uniformities in the metal as well as environmental non uniformities, such as those created by differential aeration cells or by heavy metals plated out on the surface. Localized corrosion in a microscopic sense results from galvanic coupling and stray – current effects[21].

1-10-1 Pitting

Pitting is the most common form of localized corrosion and frequently is difficult to associate with specific metallographic features. Pit shape can vary from shallow depressions to cylindrical or roughly hemispherical cavities[21].

These shapes distinguish pitting from intergranular or exfoliation corrosion. Super purity aluminium has the highest resistance to pitting, and, among the 1xxx aluminums, resistance improves with purity.

Among commercial alloys, those of the 5xxx group have the lowest pitting probability and penetration rates, followed by alloys of the 3xxx group[21].

1-10-2 Intergranular Corrosion

Intergranular corrosion is a selective attack of grain boundaries. The mechanism is electrochemical, resulting from local cell action in the boundaries. Microconstituents precipitate in grain boundaries have a corrosion potential differing from that of adjacent solid solution and transition precipitate structure and form cells with it[21].

In alloys of the 5xxx and 7xxx groups, the precipitates (Al_8Mg_5 , MgZn , and $Al_2Mg_3Zn_3$) and anodic to the matrix. In 2xxx alloys, the precipitates (Al_2Cu and Al_2CuMg) are cathodic.

Intergranular corrosion can occur with either type. Susceptibility depends on the extent of Intergranular precipitation, which is controlled by fabricating or heat treating parameter[21].

In 2xxx alloys, grain – boundary precipitation is caused by an inadequate cooling rate during the quenching operation of heat treatment.

Alloys of the 6xxx group with a balanced magnesium silicon ratio (Mg_2Si proportions) show little tendency toward Intergranular corrosion; susceptibility is higher in those with excess silicon over the Mg_2Si ratio[21].

Because Intergranular corrosion is involved in SCC of aluminium alloys, it is often presumed to be more deleterious than pitting or general (uniform) corrosion. However, in alloys that are not susceptible to SCC – for example, the 6xxx series alloys – Intergranular corrosion is usually no more severe than pitting corrosion, tends to decrease with time, and, for equal depth of corrosion, the effect on strengths is no greater than that of pitting corrosion, although fatigue cracks can be more likely to initiate at areas of Intergranular corrosion than at random pits[21].

1-10-3 Exfoliation Corrosion

Exfoliation corrosion is selective attack that proceeds along multiple subsurface paths roughly parallel to the surface. It can be Intergranular but also is associated with striated insoluble microconstituents and dispersion bands aligned parallel to the product surface[21].

It is most common in thin – section products with highly worked, flattened, and elongated metallurgical structures. Leafing or delamination accompanied by swelling caused by apparent in metallographic section.

Exfoliation frequently proceeds from sheared edges and may be initiated at pit surfaces. It is not accelerated by applied stress but is intensified by slightly acidic solutions and by galvanic coupling.

In 2xxx and copper – containing 7xxx alloys, exfoliation is considerably affected by section thickness and corresponding microstructure and by temper[21].

In the copper – containing 7xxx group alloys, resistance is greatly improved by averaging beyond peal strength.

In extrusions of these heat – treated alloys, the recrystallized peripheral zone near surfaces is often highly resistant to exfoliation while the underlying unrecrystallized portion may be vulnerable to this type of attack[21].

The 1xxx aluminium and 3xxx alloys are highly resistant to exfoliation in all tempers. Exfoliation has been encountered in highly cold worked, high – magnesium 5xxx alloys such as 5456 boat hull plate[21].

1-10-4 Galvanic Corrosion

Most forms of corrosion discussed previously are electrochemical in nature and involve cells formed by microstructural or environmental features. A number of other conditions establish potential differences that intensity and localize corrosion[21].

The accelerated corrosion resulting from electrical contact with a more – noble metal or with a nonmetallic conductor such as graphite is termed "galvanic corrosion". The most common examples occur when aluminium alloys are joined to steel or copper and are exposed to wet saline environments. In such situations the aluminium is more rapidly corroded than it would be in the absence of the dissimilar metal[21].

For each environment, metals can be arranged in a galvanic series from most to least active. Table (1-3) listed potentials based on measurements in sodium chloride solution.

The rate of corrosive attack when two metals are coupled depends on several factors :

a- the potential difference.

b- the electrical resistance between the metals.

c- the conductivity of the electrolyte.

d- the cathode / anode area ratio.

e- the polarization characteristics of the metals.

Although the corrosion potential can be used to predict which metal will be attacked galvanically, the extent of attack cannot be predicted because of polarization. For example, the potential difference

36

between aluminium and stainless steel exceeds that between aluminium and copper[21].

In natural environments, including saline solutions, zinc is anodic to aluminium and corrodes preferentially, giving protection to the aluminium.

Magnesium is also protective, although in severe marine environments, it can cause corrosion of aluminium because of an alkaline reaction. Cadmium is neutral to aluminium and can be used safely in contact with it[21].

Copper and copper alloys, brass, bronze, and monel are the most harmful followed closely by carbon steel in saline environments. Nickel is less aggressive than copper, approaching stainless steel in effect, as does chromium electroplate. Lead can be used with aluminium, except in marine environments[21].

Table (1-3) : *Electrode potentials of representative aluminium alloys and other metals.*

Al alloy or other metal	Potential* (V)	Al alloy or other metal	Potential* (V)
Chromium	+0.18 to -0.4	Mild carbon steel	-0.58
Nickel	-0.07	5052, 5086	-0.85
Silver	-0.08	5454	-0.86
Stainless steel (300 series)	-0.09	5456, 5083	-0.87
Copper	-0.20	7072	-0.96
Tin	-0.49	Zinc	-1.10
Lead	-0.55	Magnesium	-1.73

* measured in an aqueous solution of 53g of sodium chloride and 3g of H_2O_2 per liter at 25°C; 0.1N calomel reference electrode[21].

1-10-5 Stray – Current Corrosion

Whenever an electric current is conducted from aluminium to an environment such as water, soil, or concrete, the aluminium is corroded in the area of anodic reaction in proportion to the current. At low current densities, the corrosion may be in the form of pitting, while at higher at rates that do not diminish with time[21].

In soils, stray – current corrosion can be caused by close proximity to other buried metal systems, which are protected by an impressed – current cathodic – protection system. The ground current can leak onto a buried aluminium structure at one point, then off at another where the corrosion occurs, taking a lower – resistance path between the driven buried anode and the nearby structure being protected. Common bonding of all buried metal systems in close proximity is the usual way to avoid such attack[21].

1-10-6 Deposition Corrosion

Deposition corrosion is a special form of galvanic corrosion that causes pitting. It occurs when particles of a more – cathodic metal plate out of solution on the aluminium surface, setting up local galvanic cells.

The ions aggressive to aluminium are copper, lead, mercury, nickel and tin, often referred to as heavy metals. Their effects are greater in acidic solutions their solubility is low[21].

Copper ions most commonly cause this type of corrosion in applications of aluminium. For example, rain runoff from copper roof flashing can cause corrosion of aluminium gutters with no electrical contact between the two metals. Very small amounts of copper in solution (as low as 0.05ppm) can be detrimental[21].

The inferior general corrosion resistance of alloys containing copper is attributed to deposition corrosion by copper replated from the dissolved corrosion products. Mercury is the ion most aggressive to aluminium, and even traces can cause serious problems[21].

Liquid mercury does not wet aluminium, but if the natural oxide film on the aluminium surface is broken, aluminium dissolves in the mercury, forming amalgam, and the corrosion reaction becomes catastrophic.

In corrosive solutions, any concentration of mercury greater than a few parts per billion should be cause for concern.

1-10-7 Crevice Corrosion

If an electrolyte is present in a crevice formed between two facing aluminium surfaces or between an aluminium surface and a nonmetallic material such as a gasket, localized corrosion in the form of pits or etch patches can occur. This is the result of formation of a concentration cell or differential aeration cell[21].

Staining that occurs on inter – wrap surfaces of coiled sheet or foil or in packages of flat sheet or circles is a result of the same mechanism and can be preliminary to more severe corrosion that will make separation difficult. Such damage can be prevented by ensuring that the product is initially dry and by avoiding ingress of moisture by protecting it against condensation, rain, and other sources of contamination[21].

1-10-8 Filiform Corrosion

Sometimes termed "worm track" corrosion, occurs on aluminium when it is coated with an organic coating and exposed to warm, humid atmospheres. The corrosion appears as threadlike filaments that initiate at defects in the organic coating, are activated by chlorides, and grow along the metal / coating interface at rates to 1mm/day (0.04 in./day). The moving end of the filament is called the tail.

The occurrence of Filiform corrosion on painted surfaces in aircraft exposed to marine and other high – humidity environments has been controlled by use of chemical conversion coatings, anodizing, or application of chromate – inhibited primers prior to find coating[21].

1-10-9 Stress – Corrosion Cracking

Time – dependent cracking under the combined influence of sustained tensile stress and a corrosive environment is labeled SCC. In aluminium products, SCC, which is characteristically Intergranular in nature, has been experienced only in higher – strength alloys and tempers of the 2xxx, 7xxx, and 5xxx type (with more than 3% Mg) and of the 6xxx type (with excess silicon)[21].

No SCC problems have been encountered in service with 1xxx aluminium or with 3xxx, 6xxx (Mg_2Si ratio) or 5xxx (containing 3% Mg or less) alloys. In general, high – tensile stress is a prerequisite to cracking when the stress direction is parallel to either the longitudinal or the long – transverse direction. When the tensile stress is in the short – transverse direction (perpendicular to the surfaces of plate or across the flash plane of die forgings), SCC can occur in susceptible alloy / temper combinations at relatively low stresses[21].

Cracking is accelerated by aggressive, chloride – containing environments, but can occur in humid air. Because of the orientation – dependence of SCC, it is important to minimize stresses in the most susceptible direction. In addition to the stresses imposed by service loading, the residual stresses from quenching or forming, and any resulting from interference fits or assembly misfits, must be taken into account[21].

Minimizing these stresses in the short – transverse direction greatly reduces the probability of SCC failure of susceptible alloy / temper

40

combinations. Medium – strength copper – free and low – copper alloys of the 7xxx group tend to be susceptible to SCC. Successful use of 7039, which has poor resistance, in armor – plate applications requires control of short – transverse stresses and weld overlays. Alloys 7016, 7021, and 7029 with copper contents up to 1% have good formability and finishing properties for automotive applications such as bumpers[21].

1-10-10 Corrosion Fatigue

Fatigue strengths of aluminium alloys are lower in such corrosive environments as seawater and other salt solutions than in air, especially when evaluated by low – stress, long – duration tests. Such corrosive environments produce smaller reductions in fatigue strength in the more corrosion – resistant alloys, such as the 5xxx and 6xx series, than in the less resistant alloys, such as 2xxx and 7xxx series[21].

Like SCC, corrosion fatigue requires the presence of water. In contrast to SCC, however, corrosion fatigue is not appreciably affected by test direction, because the fracture that results from this type of attack is predominantly transgranular.

1-10-11 Hydrogen Embrittlement

Only recently has it been determined that hydrogen embrittles aluminium. For many years, all environment cracking of aluminium and aluminium alloys was represented as SCC; however, testing in specific hydrogen environments has revealed that aluminium is susceptible to hydrogen damage[21].

Hydrogen damage in aluminium alloys may take the form of Intergranular or transgranular cracking or blistering. Blistering is most often associated with the melting or heat treatment of aluminium, in which reaction with water vapor produces hydrogen. Blistering due to hydrogen is frequently associated with grain – boundary precipitates or the formation of small voids. Blister formation is different from that in ferrous alloys in that in aluminium it is more common to have a multitude of near – surface voids that coalesce to produce a large blister[21].

In a manner similar to the mechanism in iron – base alloys, hydrogen diffuses into the aluminium lattice and collects at internal defects.

This occurs most frequently during annealing or solution treating in air furnaces prior to age hardening. Most of the work on hydrogen embrittlement of aluminium has been on the 7xxx alloys; therefore, the full extent of hydrogen damage in aluminium alloys has not been determined and the mechanisms have not been established[21].

1-10-12 Erosion – Corrosion

In noncorrosive environments, such as high – purity water, the stronger aluminium alloys have the greatest resistance to erosion – corrosion because resistance is controlled almost entirely by the mechanical components of the system. In a corrosive environment, such as seawater, the corrosion component becomes the control lings factor; thus, resistance may be greater for the more corrosion – resistant alloys even though they are lower in strength[21].

1-11 The Literature Survey
1-11-1 General Corrosion

The electrochemical behaviour of aluminium and its alloys had attracted many investigator's attention due to it's great importance.

Stella (1978)[36] studied the electrochemical behaviour of pure Al in different (Cl⁻) media similar to those formed inside growing pits.

Artificial pit measurements confirm the previous assumption that saturated AlCl₃ must be formed at the pit bottom as a condition for the attack. Experimental results support the idea that a competitive process involving passivation and activation can occur depending on the potential, pH and (Cl⁻) concentration.

Nisancioglu and Holtan (1978)[37] measured of the critical pitting potential of aluminium and its alloys in chloride media by using current controlled, potential controlled, and open circuit methods.

Also they studied[38] the corrosion behaviour of aluminium in chloride solutions at potentials below the critical pitting potential by using potentiodynamic method and steady state polarization data.

Zaki (1981)[39] studied corrosion behaviour and corrosion prevention of aluminium alloys in desalination plants by electrochemical techniques and particularly the polarization resistance method who was observed :

1- The Al –Mg alloys showed the best resistance to pitting.

2- Alloys 1199, 5454, and 5457 showed resistance to pitting although to a lesser degree. Fouling did not appear to increase the rate of corrosion of these alloys significantly.

3- Alloys 5083, 5457, 5454, and 5154 showed relatively more resistance to corrosion.

The reactions of aluminium were:

$$Al \rightarrow Al^{+++} + 3e$$
$$2Al + 3H_2O \rightarrow Al_2O_3 + 6H^+ + 6e \quad \bigg\} (Anodic)$$
$$Al + 2H_2O \rightarrow AlO_2^- + 4H^+ + 3e$$

and

$$Al^{+++} + 3e \rightarrow Al\downarrow$$

$$Al\downarrow + 3/4\,O_2 \rightarrow 1/2\,Al_2O_3 \quad ;T>82.2^oC$$

$$Al\downarrow + 2H_2O \rightarrow AlOOH + 3/2\,H_2 \; ; T<82.2^oC$$

(Cathodic)

Also he studied[40] the effect of velocity, dissolved oxygen, pH control, chlorination, temperature, and dissolved ions on galvanic corrosion of aluminium alloys (1100, 3003, 5052, 5054, and 6061). He observed that the reaction of chlorine are simple as given below :

$$Cl_2 + H_2O \rightarrow 2H^+ + 2Cl^- + 1/2O_2$$

$$Cl_2 + H_2O \rightarrow H^+ + Cl^- + HOCl$$

The effect of chromate and carbonates among other dissolved ions was more effective.

Salvarezz, Mele, and Videla (1981)[41] studied the redox potential and the microbiological corrosion of aluminium and its alloys in fuel/water systems, where the association of different species of bacteria and fung, the use of this parameter is of special interest to determine the importance of each specie in the corrosion process.

Cune, Shilts, and Ferguson (1982)[42] studied the growth of hydrated surface films on 99.99% pure aluminium in aqueous media at 100°C with Rutherford backscattering spectroscopy (RBS). Also they studied the effect of the inhibitors anions borate, molybdate, nitrate, phosphate, sulfate, silicate and tungstate, and of ethylene glycol on this reaction at concentration levels typical of automotive coolant formulations and they classified the behaviour of the inorganic anions into three types.

Mauret and Lacaze (1982)[43] studied the water corrosion of Al-Mg (5154) and Al-Cu-Mg (2024) foils based on gas chromatography.

Measurements of volumes of hydrogen evolved by the attack of distilled water on the alloys :

$Al + 3H_2O \rightarrow Al(OH)_3 + 3/2\ H_2$

$Mg + 2H_2O \rightarrow Mg(OH)_2 + H_2$

at 70°C in a variable frequency oscillating oven.

Sanad, Ismail, El – Sobki and Shalaby (1982)[44] studied the corrosion behaviour of aluminium and Al – Mg alloys by means of weight – loss measurements and potentiokinetic experiments at different temperature (30, 50, 70 and 100°C) in 32.7 g/L sodium chloride.

The corrosion – resistant alloys were found to be 5%, 1.5% and 3% Mg at a temperature 30, 70 and 100°C respectively. The alloying of Al with Mg had no effect on the pitting potential till 70°C. At 100°C the pitting potential of pure Al became more negative.

Schwabe, Herrmann and Berthold (1983)[45] studied the anodic current potential curves of Al(99.999%) in H_2O/1M H_2SO_4 are quite different from that in anhydrous DMF/1M H_2SO_4. They observed that Al is active in DMF/H_2SO_4, and at potential of 0.4V (SCE) give a constant current – time curve, where in H_2O/H_2SO_4 the current decreases in the same time to a low value. Comparison of cut – off curves after polarization to 0.4 and 3V (SCE) show that the curves are similar, but the potential decreases more slowly in DMF/H_2SO_4 than in H_2O/H_2SO_4.

Bogustaw (1983)[46] studied the electrochemical behaviour of Al – Mg alloys. Stationary potentiodynamic and potentiodynamic studies were performed on synthetically obtained Al_8Mg_5 samples. A tendency for passivity in electrolytes of pH value between 4 and 10 in a limited

45

potential range. For high pH values the Al_8Mg_5 compound was passive over the entire range of anodic potentials. In the corrosion and anodic polarization of Al_8Mg_5 magnesium dissolution is favoured rather than the formation of Al^{+3} and Mg^{+2} in stoichiometrically equivalent amounts.

The electrochemical behaviour of Al – Cu alloys was studied by Boguslaw and Antoni (1983)[47]. Stationary potentiodynamic polarization on static electrodes and rotating ring disc – electrodes and potentiodynamic polarization were carried out on an Al_2Cu compound in sulphate solutions of different pH values. Al_2Cu under self corrosion conditions as well as under anodic polarization conditions dissolves with the formation of aluminium and copper ions.

$$Al_2Cu \rightarrow Cu^{\cdot} + 2Al^{3+} + 6e$$
$$Al_2Cu \rightarrow 2Al^{3+} + Cu^{2+} + 8e \text{ , and } Cu^{2+} + 2e \rightarrow Cu^{\cdot}$$

Csanady and Co – workers (1984)[48] studied the relationship between the corrosion resistance and impurity content of aluminium oxide layers. In this study it has been shown that the aluminium – water vapour reaction is greatly affected by the composition and structure of the amorphous oxide layer on the surface and this depends on the composion of base metal and on the annealing of the foil. In particular the corrosion resistance is decreased by presence of alkali (Li) and alkaline earth metal (Mg) cations in the oxide layer.

Hunkeler and Bohni (1984)[49] studied the mechanism of pit growth on aluminium under open circuit conditions and they compared with the results under potentiostatic conditions. They concluded that pit growth is ohmically controlled in both cases, which under ideal conditions results in a square root growth law. Also they discussed the influences of various factors such as the ohmic potential drop outside of the pit, the number of pits, time dependent electrolyte conductivity in the pit, the Tafel constant, and the cathodic current on the growth law.

Michael, Chen, and Shirn (1984)[50] studied the electrochemical behaviour of Al in HCl by cyclic voltammetry, square – wave chronoamperometry, and square – wave chronopotentiometry. They studied the effect of high speed cyclic voltammetry, temperature, HCl concentration, potential sweep rate, and cathodic potential limit.

Relevant parameters that can be determined from the cyclic voltammetric $i - V$ curve include the breakdown and protection potentials for Al, the onset potential for hydrogen evolution, and the magnitude of the anodic and cathodic currents.

Abd – El – Nabey, Khalil, and Khamis (1985)[51] studied the corrosion behaviour of aluminium metal in water – organic solvent mixtures containing HCl using thermometric, hydrogen evolution, weight loss methods and electrochemically using a Tafel extrapolation method. The composition of the medium was changed from 0.0 to 60% (v/v) of each of methanol, ethanol, isopropanol and ethylene glycol.

Addition of small amounts of the alcohol to the aqueous medium produced a marked reduction in the corrosion current. In general, the corrosion rate decreases with increasing percentage of alcohol up to 20%, after which further increase in the range 20 – 60% has little effect on the corrosion current. They interpreted the result in term of the dependence of corrosion rate on the structural properties of water – alcohol mixtures.

Onuchukwu and Oppong – Boachie (1986)[52] studied the corrosion characteristics of 1060 Al alloy in different concentration of *p-* quinone and acetic acid in 6.5% KNO_3 as supporting electrolyte by a potentiostatic polarization technique. The suggested reactions were :

$Al^{\bullet}{}_{(s)} + H_2O_{(l)} \leftrightarrow AlOH_{(ad)} + H^{+} + e$

$2AlOH_{(ad)} + H_2O_{(l)} \rightarrow Al_2O_{3(s)} + 4H^{+} + 4e$

$AlOH_{(ad)} + 2CH_3COOH \rightarrow AlO.2[O.COCH_3] + 3H^{+} + 3e$

$$O={\bigcirc}=O \overset{+2e}{\leftrightarrow} \text{-}O\text{-}{\bigcirc}\text{-}O\text{-} \overset{2H^{+}}{\leftrightarrow} HO\text{-}{\bigcirc}\text{-}OH$$

    Quinone   Benzoid ion   Hydroquinone

Zaki (1986)[53] studied the kinetics of anodic and cathodic polarization of aluminium and its alloys. He discussed the three region of typical pitting scan of modified aluminium alloy 2778 in Arabian Gulf water. He showed that the anodic reaction takes place nearly on all the surfaces, whereas the cathodic reaction takes place only on impurities and grain boundaries. The cathodic reaction stops as soon as the cathodic sites are destroyed and the film achieves an equilibrium thickness.

The main anodic reaction :

$Al + 3H_2O \rightarrow Al(OH)_3 + 3H^{+} + 3e$

$Al + 2H_2O \rightarrow AlOOH + 3H^{+} + 3e$

$Al + 3/2H_2O \rightarrow 1/2Al_2O_3 + 3H^{+} + 3e$

Also he studied the effect of alloying elements, the following order for some binary aluminium alloys in NaCl was obtained in terms of decreasing pitting potential from the anodic polarization studied :

Al-Cu > Al-Ni > Al-Si > Al-Mn > Al-Cr > Al-B > Al-Zr > Al-Ti > Al-Mg > Al-Be

Cabot, Garrido, Perez, and Virgili (1986)[54] studied the cathodic polarization of an oxidized 99.9995% aluminium electrode at potentials from -2 to 2.5 V(SCE) in several electrolytes of neutral and acid pH values by potentiodynamic oxidation of cathodically polarized aluminium electrodes following the technique of Rozenfel'd et. al. The cathodic behaviour in acid and neutral media is interpreted on the basis of a local alkalization resulting from the H_2 evolution.

Moshier, Davis, and Ahearn (1987)[55] studied the corrosion and passivity of aluminium. The composition and thickness of the oxide/hydroxide film that forms on pure aluminium surfaces that are polarized in 0.05M Na_2SO_4 in acidic, near – neutral, and alkaline solutions using X – ray photoelectron spectroscopy (XPS). Although a thin layer of gibbsite [$Al(OH)_3$] is often indicated to be the film that forms and protects aluminium from corroding in near – neutral pH solution after 10 – 20h of exposure, the compositions of the film that initially forms on Al after a short period of time (1 h) was actually closer to boehimite (AlOOH), the oxyhydroxide phase. The growth of the film is driven by electrochemical reactions and the reason behind the increase in corrosion is that the formation of the hydroxide :

$$Al + 3H_2O \rightarrow Al(OH)_3 + 3H^+ + 3e$$

At neutral pH is replaced by the dissolution of Al

$$Al \rightarrow Al^{3+} + 3e$$

at low pH or the formation and dissolution of the film

$$Al + 3OH^- \rightarrow Al(OH)_3 + 3e \ , \qquad Al(OH)_3 + OH^- \rightarrow Al(OH)_4^-$$

in alkaline conditions.

Lunarska and Szklarska (1987)[56] studied the changes in morphology and chemical composition of corrosion on the surface of three powder metallurgy AlZnMg alloys containing 8.8 to 12.5% Zn, 2.4 to 2.5% Mg and 1.2 to 1.5Cu after their exposure to deaerated 3.5% NaCl solutions of pH 1, 7 and 13 at room temperature in a wide range of applied potentials, using scanning electron microscopy (SEM) and energy dispersive spectroscopy (EDS) techniques.

Also they studied[57] the stress corrosion cracking (SCC) of three wrought powder metallurgy (P/M) AlZnMg alloys differing in Zn content and heat – treatment in deaerated 3.5% NaCl solution over a wide range of applied potentials, using the slow strain rate tensile technique (SSRT).

Mazhar, Heakal, and Mogoda (1987)[58] studied the dissolution behaviour of anodic oxide films formed on aluminium in phosphoric acid solutions by impedance and potential measurements. The dissolution follows a first order mechanism and the dissolution of the outer layer increases with increasing acid concentration, while the inner layer is not affected in the same degree by changes in acid concentration.

The influence of some anions on the rate of dissolution of the oxide indicates that among PO_4^{3-}, SO_4^{2-}, NO_3^-, Cl^-, Br^-, and ClO_4^-, the last two are the most aggressive.

Qvari, Tomcsanyi, and Turmezey (1988)[59] discussed the electrochemical study of the pitting corrosion of aluminium and its alloys and they determined the critical pitting and protection potentials by potentiostatic method. They studied the adsorption and absorption of Cl^- ions on the working electrode.

Surganov, Jansson, Nielsen, and Morgen (1988)[60] studied the different stages of aluminium anodization in oxalic acid solution by noble gas sputter sectioning combined with Auger electron spectroscopy (AES) and Rutherford backscattering spectroscopy (RBS).

Hong – pyo and Co – workers (1988)[61] studied stress cosrrosion cracking (SCC) and corrosion behaviour of Al – Zn – Mg alloys in deaerated 3.5wt% NaCl solution at pH = 3.5 with and without the hydrogen recombination poisons (HRPs) $Na_2S.9H_2O$ or Ca_3P_2.

Trujino and Oki (1988)[62] studied the aluminium corrosion rates in aqueous solutions using a galvanic couple reaction. The current through a galvanic couple using a zero impedance ammeter had an extremely high correlation with the corrosion rate obtained from weight loss measurement.

50

Tan and Chin (1989)[63] studied the effect of 60 – Hz sinusoidal, square, and triangular alternating voltages (AV) on the corrosion of aluminium in dilute nitric acid, sodium nitrate, sodium sulfate, and sodium chloride solutions.

Tomcsanyi and Co – workers (1989)[64] studied the composition of the passive layer on aluminium and reformation of this layer during cathodic polarization by radiotracer and electrochemical methods in the presence of sulphate and chloride ions.

They discussed formation of hexacoordinated aluminium ion which forms different hydroxo complexes in neutral solution (pH = 4-10) according to the following:

$$[Al(H_2O)_6]^{3+} \rightarrow \begin{cases} [Al\,(OH)(H_2O)_5]^{2+} \;+\; H^+ \\ [Al\,(OH)_2(H_2O)_4]^+ \;+\; 2\,H^+ \\ [Al\,(OH)_3(H_2O)_3] \;+\; 3H^+ \\ [Al\,(OH)_4(H_2O)_2]^- \;+\; 4\,H^+ \end{cases}$$

O^{2-} ions are also formed in the solid phase by topochemical reaction to produce $[Al\,O_x(OH)_y(H_2O)_z]$ and then transforms in aqueous media according to the following scheme :

$$Al[\,O_x(OH)_y(H_2O)_z] \rightarrow (AlOOH)_4*H_2O \rightarrow AlOOH \xrightarrow{+\;H_2O} Al(OH)_3$$
Gelatinous alumina   Pseudoboehmite   Boehmite   Bayerite

Chloride ion is bonded chemically in the interface as an initial step of the formation of different mixed oxohydroxo – and chloro complexes according to the following formula :

$Al[O_x(OH)_y(H_2O)_z] + Cl^- \rightarrow Al[O_x(OH)_{y-1}\,Cl(H_2O)_z] + OH^-$

$(AlOOH)_4*H_2O + Cl^- \rightarrow (AlOOH)_3*AlOCl*H_2O + OH^-$

$AlOOH + Cl^- \rightarrow AlOCl + OH^-$

$Al(OH)_3 + Cl^- \rightarrow Al(OH)_2Cl + OH^-$

Finally the $[AlCl_6]^{3-}$ complex is produced.

51

Furuya and Soga (1990)[65] measured the critical pitting potential of aluminium alloys sold on the market in a solution featuring a chloride ion concentration of 0.01 to 1.0 M/l at 25°C.

Also they used the Cu^{++} ion method for measuring the critical potential, and a high degree of correspondence was shown with the potentiodynamic method hitherto in use.

Elboujdaini and Ghali (1990)[66] studied the anodic behaviour of aluminium alloys (5083 and 6061) in aqueous chloride solution in the presence of sulphate ions by the linear sweep polarization technique and potential step method. Significant pitting was observed for 5083, as a result of the combined effects of the electrochemically active Mg in the matrix of the electrode, its subsequent influence on hydrogen evolution, and the chloride attack on the aluminium electrode surface.

Xue and Co – workers (1991)[67] investigated the effect of fluoride ions on the corrosion of aluminium in sulphuric acid through thermodynamic analysis and corrosion experiments. Study of the solution chemistry of aluminium in aqueous solution in the absence and in the presence of fluoride ions with the construction of Eh – pH diagrams for the $Al - F - H_2O$ systems at 25°C.

In the presence of fluoride ions, the corrosion of pure aluminium in sulphuric acid is due to uniform dissolution and the reaction rate depends on the fluoride concentration according to following equations :

$Al_2O_3 + (2n) HF + (6-2n) H^+ \leftrightarrow 2AlF_n^{(3-n)+} + 3H_2O$

For the oxide film, and the reaction

$Al + nHF + (3-n)H^+ \leftrightarrow AlF_n^{(3-n)+} + 3/2 H_2$ for the metal.

Holzer and Co – workers (1993)[68] studied the anodic oxidation of various aluminium alloys by means of rotating disc electrodes in 3M H_2SO_4 as function of Cl^-, F^-, Zn^{2+} and In^{3+} concentration.

Saidman, Garcia, and Bessone (1995)[69] studied the electrochemical behaviour of aluminium in NaCl solutions containing In^{3+} ions and the dissolution of Al – In and In – Al alloys using potentiostatic, galvanostatic and potentiodynamic techniques, complemented by SEM.

William and Co – workers (2002)[70] studied the effect of chromate treatment on the corrosion of an Al/Cu aircraft alloy with a galvanic corrosion aperture composed of two electrodes and a zero resistance ammeter. Combination of pure Al, pure Cu, and 2024 electrodes were immersed in 0.1M NaCl solution, which was saturated with air, O_2, or argon.

1-11-2 Corrosion Inhibition

There are many investigators tried to control on corrosion of aluminium and its alloys by using many methods specially the inhibitors.

Desai and Co – workers (1976)[71] used aliphatic polyamines and ethanolamine as corrosion inhibitors for Al – 515 in HCl solution.

The performance of inhibitors under the influence of external cathodic polarization.

In (1980) Rudd and Scully[72] measured the pitting potential of aluminium in aqueous solutions containing 3×10^3 ppm Cl^- ion and one of six inhibitors in the range 0.08 – 0.32M, over the pH range 4-9 and in glycol/water mixture. The inhibitors were effective in raising the pitting potential in decreasing order :

nitrate >> phosphate > citrate = tartrate > benzoate > acetate.

While Talati, A. Patel, and P. Patel (1980)[73] studied the corrosion of aluminium alloy (99.8% Al), (Al – 1.3%Mn), (Al – 3.9% Cu) and (Al – 2.2% Mg) and their inhibitors (aminophenols) in solutions of H_3PO_4.

Abo El – Khair and Ateya (1981)[74] studied the inhibiting effect of triphenyl tetrazolium chloride (TTC) on the corrosion af aluminium in 1N HCl.

Singh and Agarwal (1982)[75] tested sodium chromate and sodium hydrogen phosphate separately, as well as in combined state, for inhibiting the corrosion of some commercially – pure aluminium alloys (1060, 1100, 3003, and 5052) in various concentrations of HNO_3.

In (1983) Talati and Gandhi [76] studied the inhibition of the corrosion of Al – 4Cu alloy in solution of HCl by some N – hetrocyclic compounds in relation to the concentration of the acid and of inhibitor as well as temperature.

While Abo El – Kair and Mostafa(1983)[77] studied the inhibitive effect of N – vinylpyrrolidone and polyvinylpyrrolidone of three different molecular weight on the dissolution of Al in 1.15N HCl.

But Yadav, Chandhary and Agarwall (1983)[78] studied the inhibitive effect of aliphatic amines on the corrosion of Al alloy (1060, 1100 and 3003) in an acidic chloride solution of pH=1 in the presence and absence of tungstate ions.

Also Chakrabart, Singh and Agarwal (1983)[79] demonstrated the corrosion inhibiting efficiencies of guanidine carbonate and some of its condensation products for aluminium (1060) in 20% HNO_3 at 25°C and 35°C.

In (1984) Talati, Patel, and Gandhi[80] studied the cathodic protection of Al alloy 2017 (Al – 4%Cu) in 2.5M HCl containing various dyes and of 2017 and 3003 (Al − 1.3% Mn) in 2.0M trichloroacetic acid (TCA) containing various amines as inhibitors.

While Abo El − Khair and Mostafa (1984)[81] studied the effect of polyfunctional group compounds (ethylene diamine, ethylendiaminetetra acetic acid, nitro − triacetic acid, citric acid) on the corrosion of aluminium in 1N HCl at various inhibitor concentration.

El Sayed (1984)[82] studied the effect of some hydrazine and hydrazone derivatives as inhibitors for the dissolution of aluminium in 2M NaOH.

Also Dossoki and Co − workers (1984)[83] studied the efficiency of some anthraquinone derivatives in retarding the dissolution of aluminium in HCl and NaOH media.

Fouda and Co − workers (1986)[84] investigated the inhibitive action of some thiosemicarbazide derivatives towards the corrosion of aluminium in 2M HCl by using thermometric, weight loss and hydrogen evolution techniques.

By same techniques Ahmed and Co − workers (1988)[85] studied the inhibitive effect of some morpholine and thiosemicarbazide derivatives on the dissolution of aluminium in 2M HCl at various inhibitor concentrations and temperature.

Arnott, Hinton, and Ryan (1989)[86] studied the Cationic − film − formation inhibitors for the protection of the 7075 aluminium alloy against corrosion in 0.1M NaCl solution. They observed decreasing the corrosion rate by the addition of 1000ppm of $FeCl_2$, $CoCl_2$, $NiCl_2$, YCl_3, $LaCl_3$, $NdCl_3$, $PrCl_3$, or $CeCl_3$.

In the same year Tagouri and Mostafa[87] studied the efficiency of some 2- hetrocarboxaldehyde − 2' − ptridyl − hydrazones as inhibitors for the dissolution of Al in 2M HCl at different temperature and concentrations, maximum percentage protection (85%) was obtained at 10^{-3}M for the compound containing oxygen atom.

In (1991) Ahmed and Co – workers[88] studied the effect of tosyl – hydrazine, 4- nitrobenzoyl – hydrazine, and terepthalyl – hydrazine on the corrosion of aluminium in H_2SO_4 by weight loss and polarization measurements.

Radosevic and Kliskic (1992)[89] attempted to inhibit cathodic corrosion of aluminium in chloride solution by the use of organic acids and bases. Hence, organic substances were expected to inhibit the corrosion by consuming OH⁻ ions as well as by an adsorptive effect.

In (1994) Metikos and Co – workers[90] determined the inhibiting effects of 1 and 2- naphthylamines on the corrosion kinetics of aluminium in acid media ($HClO_4$ and $NaCl + HCl$) using electrochemical methods.

The action of inhibition is due to the planar orientation (with π – electron bonds) of adsorbed inhibitor molecules and the existence of synergetic effects.

Frankel and Richard (2001)[91] studied the mechanism of inhibition Al alloys by chromate conversion coating (CCC). They discussed the polarization curves for 2024 in O_2 bubble 1M NaCl containing various amounts of $Na_2Cr_2O_7$. Also they illustrated the polymerization of Cr^{III} following reduction of Cr^{VI} during CCC formation.

Wood and Co – workers (2002)[92] studied the pitting corrosion inhibition of aluminium 2024 by Bacillus biofilms secreting polyaspartate or γ – polyglutamate.

Luis and Co – workers (2003)[93] investigated the cerium conversion layer (CeCl) as a replacement for chromium conversion layer to protect Al alloys against corrosion. In this work the microstructure and the electrochemical behaviour of aluminium alloy 2024 with and without CeCl were investigated and the results have shown that the presence of dispersed plated Cu particles on the alloy surface enhances the formation of the CeCl.

Results of potentiodynamic experiments have shown that the corrosion protection afforded by the conversion layer is due to the hindrance of the oxygen reduction reaction and that the pitting potential of the alloy is not change.

1-12 The Aim Of The Present Research

The subject of this research included a number of important aims which may be summarized as :

1- Investigation of the full polarization curves of pure aluminium metal and also of three aluminium base alloys (2024, 5083, and 7075) in the basic media with three values of pH (13, 11, and 9) at four temperatures over the range of (298 – 313)K. The various regions of each polarization curve, involving the cathodic and anodic Tafel sections together with the passive and transpassive regions, are to be analyzed.

2- Studying the influence of various concentrations of chloride ions on the corrosion and passive behaviour of the pure aluminium and its three alloys in the three values of pH over the same range of temperatures.

3- Studying the inhibiting effect of three various concentrations of sodium acetate and sodium chromate on the corrosion of the metal and alloys in NaOH solution (pH = 13, 11, and 9) at four temperatures in the range of (298 – 313)K.

4- The temperature dependence of the polarization in the basic media enabled the thermodynamic and kinetic aspects of the corrosion to be studied. Extensive data could be derived from the detailed analysis of each region.

Chapter Two: Experimental Part

2-1 The Experimental Set – Up

The instrument was used in this work consists of a source of potential (an electronic voltmeter) and current source[94]. The potentiostat measures the potential (V) of the test electrode under study and compares this with the reselected value (V^*) from the potential source. If there is a difference ($\delta V = V^* - V$) between the measured and the chosen potentials, the potentiostat tells its current source to send a current (i) between the auxiliary and the test electrodes. The direction and magnitude of this current is electronically chosen to keep the potential of the electrode at the desired value, i.e., to make ($\delta V = V^* - V = 0$)[95].

A potentiostat of the type PRT 10-0.5L, which was obtained from SOLE TACUSSEL with an output voltage of $\pm 10V$, out current of $\pm 500mA$ and a response time of 2-3μs was used[96] in the measurements.

The recorder was (EPL-2B)[97] potentiostatic recorder with interchangeable plug-in pre amplifier, type EPL2, which enable the working electrode current to be recorded in either linear or logarithmic coordinates. The "CORROSCREPT" used also contained a digital electronic millivoltmeter, MVN 79 type. This instrument is intended for the high accuracy measurement of potential differences from a few millivolts to some tens of volts, across sources of very high internal resistance, all organized in a particular way (Fig. 2-1)[98].

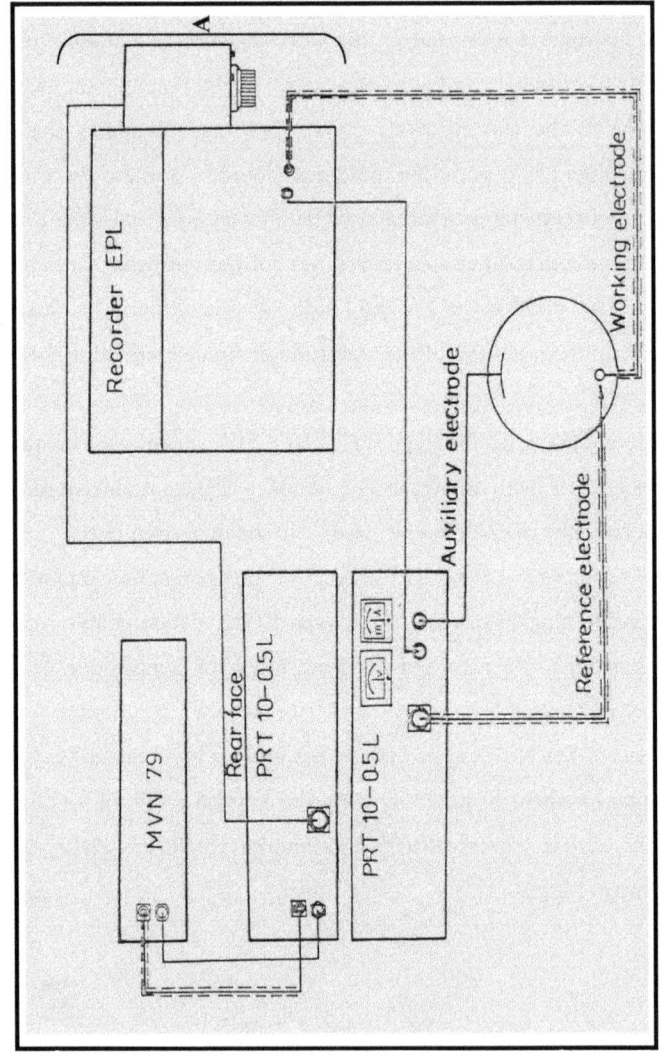

Fig. (2-1) : *The schematic diagram of the CORROSOCRIPT*

60

A simple electronic lay-out of the potentiostat is shown diagrammatically in Fig.(2-2). the potential of working electrode, E_t, is measured against another electrode E_r, called reference electrode.

A third electrode, E_a, called the auxiliary electrode allows the electrical current necessary to produce the desired potential difference to flow through the circuit[94].

The working and the auxiliary electrodes are connected to the out put terminals of the potentiostat current through the circuit, is automatically controlled so that the potential difference between the working electrode and the reference electrode takes the desired value. This process is carried out by means of a differential amplifier (A_d), one output of which e_1, to voltage source called pilot voltage (or control voltage).

The amplifier derives power (A_p), which controls the output current of the potentiostat in such a manner that the potential difference between the working electrode and the reference electrode remains equal to the applied voltage (E_c).

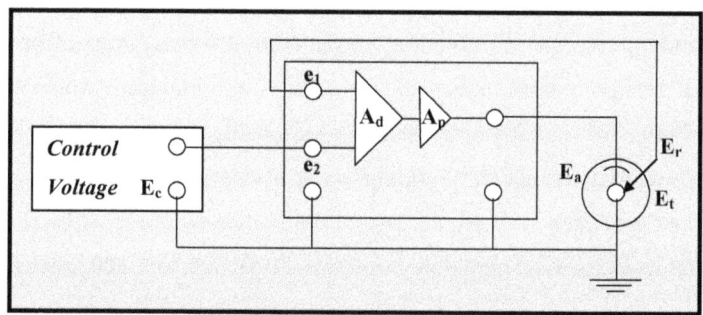

Fig. (2-2) : *The modern electronic instruments of potentiostat.*
where: E_a=Auxiliary electrode, E_r=Reference electrode,
E_t=Working electrode.

2-2 The Working Electrode[99-104]

Working electrodes of Aluminium and its alloys were made square-shape exposed surfaces of $1cm^2$. The preparation of the working electrode started by cutting the specimen using by hack saw, a band saw, a cutting-off wheel or in the sampling of hard and brittle materials, the specimen may by notched and fractured occasionally.

It was necessary to resort to flame cutting in order to obtain suitable specimens from large sections. It was essential to avoid heating and dragging the metal and the cut surface. To minimize heating during cutting, a hack saw should be used, with proper lubrication for all specimens that can be cut by this tool. Hard materials should be cut with cutting off wheels operating under water.

A practical and rapid method of mounting specimens, when harder plastics as a mounting material were used. The chief merits of these plastics were: *(1)* ability of the shape; *(2)* close contact between specimen and mounting material; *(3)* relatively low molding temperature; *(4)* inertness to the etching reagent.

Grinding consisted of abrading specimen by using a series of grits of increasing fineness, until a set of scratches was produced which were sufficiently fine to be totally eradicated in a reasonably short time by the finest polishing material available. Grinding on a abrasive papers consisted of abrading the specimen with papers coated with silicon carbide of increasing fineness. The grades most commonly used were 220, 500, and 800 mesh grit.

On passing from coarser to finer paper, it was advisable to rotate the specimen 90 degree to facility recognizing when the coarser scratches have been entirely replaced by the finer ones.

Usually, the specimens were finally polished on a soft cloth with diamond paste as a polishing material using few drops of a lubricating oil.

After that, the specimens were degreased using hot benzene and acetone. The surface area of each electrode was determined by measuring all dimensions to the nearest 0.001 mm, and subtracting the area under the Teflon gasket two-dimensional measuring microscope (pye- 6147m) was used for area measurements.

The working electrode was connected to the end of a Tungsten rod (200 mm length and 3 mm diameter). A glass container surrounded the Tungsten rod to isolate it from the solution, and light – fitting Teflon sleeve was fitted over one end of test electrode to expose an apparent surface area Fig. (2-3).

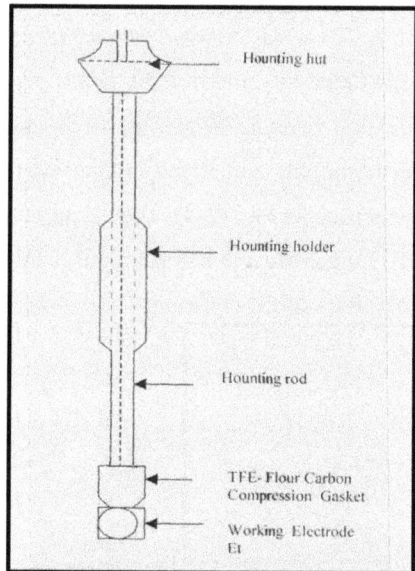

Fig. (2-3) : *A schematic diagram of working electrode.*

2-3 The Auxiliary Electrode[105-108]

The auxiliary electrode was prepared from a high purity rod stock with an exposed surface area of 1.8 cm².

Platinzed auxiliary electrode was used in the experiments due to its large surface area and high catalytic activity. Platinization of the electrode was made after cleaning the surface of platinum electrode in hot aqua regia (3 parts concentrated HCl and 1 part concentrated HNO₃), washing, and then drying. The electrode was then platinzed by immersion in solution consisting 3 percent (w/v) chloroplatinic acid and 0.02 percent (w/v) lead acetate and electrolyzing at a current density of (40 mA/cm²) for (5min). The polarity was reversed every minute. Occluded chloride was removed by electrolyzing in a dilute (10 percent) sulphuric acid solution for (5 min) with a reversal in polarity every minute. The electrode was there after raised thoroughly and stored in distilled water until ready for use in the corrosion experiments Fig. (2-4). The electrode which was obtained by this procedure had a longer life and less susceptible to poisoning due to the presence of lead acetate in its surface coating.

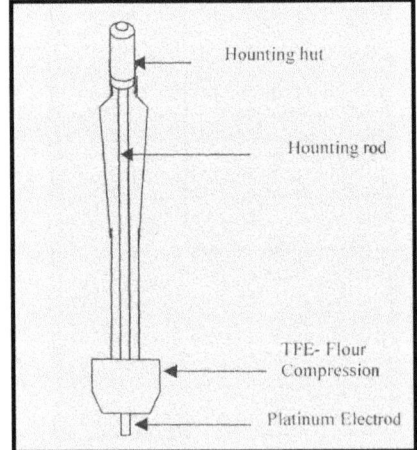

Hounting hut

Hounting rod

TFE- Flour Compression

Platinum Electrod

Fig. (2-4) : *A schematic diagram of auxiliary electrode.*

2-4 The Reference Electrode[105-108]

A saturated calomel reference electrode (SCE) was used throughout the whole work. The calomel element consisted of mercury, mercurous chloride, and chloride ion.

$$Cl^-, Hg_2Cl_{2(s)}; Hg_{(l)}$$

The reduction reaction which occurs in the calomel electrode, may be represented as :

$$Hg_2Cl_2 + 2e \longrightarrow 2Hg_{(l)} + 2Cl^-_{(aq)} \qquad \ldots\ldots(2\text{-}1)$$

The electrode is usually brought in contact with the electrolyte through a glass tubing as "Luggin capillary" which is filled by the test solution. The tip of the Luggin capillary is placed in the electrochemical cell very close to the working electrode; it was placed within (1 mm) or so of its surface – such arrangement allowed the calomel electrode to be removed entirely from the class Luggin envelope to replenish the bridge solution whenever need Fig. (2-5). Such arrangement of the Luggin capillary eliminated any IR drop that may develop in the solution separating the calomel electrode and the working electrode in the cell solution.

The calomel electrode could be prepared by grinding calomel (Hg_2Cl_2), mercury and a small quantity of saturated KCl solution together and placing the resultant slurry in a layer about (1 cm) thick on the surface of mercury contained in a clean test tube.

External contact to the mercury was usually made by a platinum wire, which was sealed , to glass.

65

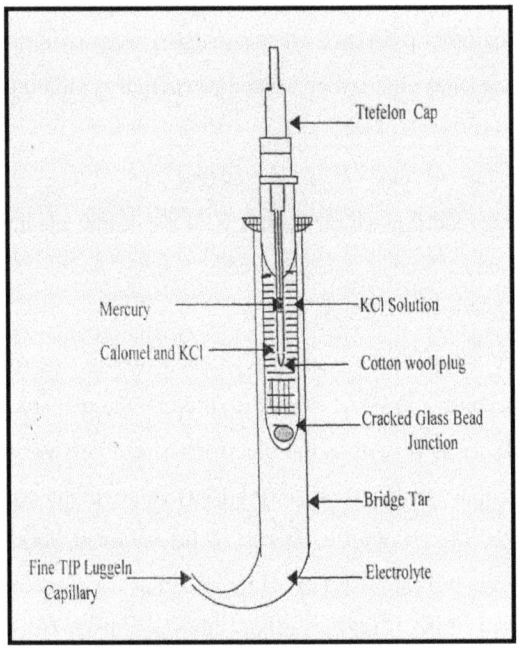

Fig. (2-5) : *A Schematic diagram of*
reference electrode.

2-5 Potentiostatic Measurement

The potentiostatic scan started about 1 hour after the electrode immersion in the test solution, beginning at about (− 2.0V) and proceeded through to (+ 2.0 V) versus the saturated calomel electrode.

The potential of the working electrode (E_t) was measured against the reference electrode (E_r) [Fig. (2-6)], while the current density was measured by ammeter which joined with the auxiliary electrode.

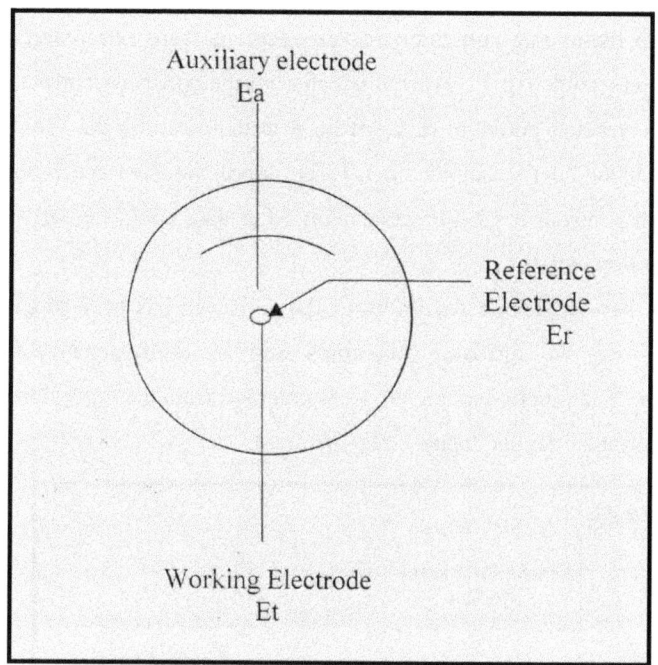

Fig. (2-6) : *A Schematic representation of a three–*
electrode cell (Corrosion cell) system.

The recorder was of EPL series potentiostatic recorder with interchangeable plug – in pre-amplifier, type EPL 2, which enabled the working electrode current density to be recorded in either linear or logarithmic coordinates.

Both the cathodic and anodic curves were obtained with decreasing and increasing polarization, and this was repeated several times.

The polarization curve obtained involved several regions covering the cathodic, anodic, passive and transpassive regions [Fig. (3-1)]. Extensive data could be derived from the detailed analysis of each polarization region. Tangents to the anodic and cathodic Tafel regions were extrapolated to the point of intersection Fig. (2-7) from which both the corrosion current density (i_{corr}) and corrosion potential (E_{corr}) were determined using the four – point method[19] cathodic (b_c) anodic (b_a) Tafel slopes, transfer coefficients (α), polarization resistances (R_p) together with other data could be derived from the polarization curves.

The thermodynamic feasibility of the corrosion has been judged from the values of the corrosion potentials and of their dependencies on temperature. The kinetic parameters were obtained from the corrosion current densities and their dependencies on temperature.

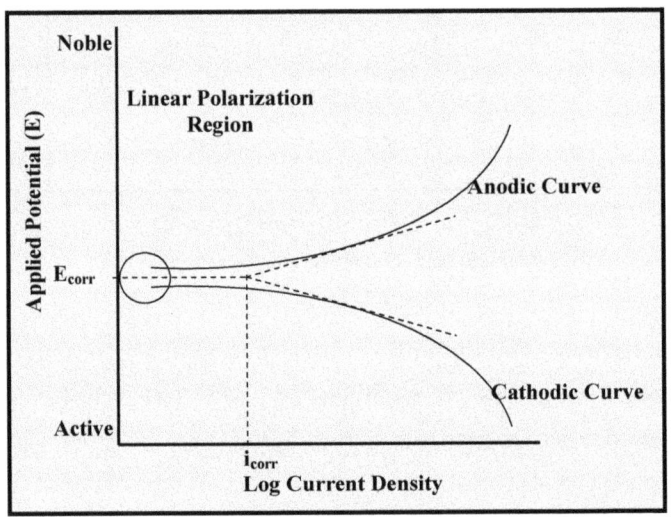

Fig. (2-7) : *Extrapolation method to determine corrosion potential and current density.*

2-6 The Experimental Techniques And Procedure

The investigation was carried out using the standard CORROSCRIPT potentiostat (TACUSSEL, France). It consisted of the following parts:

a- A transistorized potentiostat, type PRT. 10-0.5L.

b- A digital electronic millivoltmeter, type MVN 79.

c- A potentiometric recorder, type EPL – 2B.

The recorder was fitted with a plug – in amplifier, type TILOG 101, enabling currents to be plotted on either linear or logarithmic coordinates[97]. A pilot unit, type SYNCHOSCRIPT, was fitted on the right hand side panel of the recorder. This unit was basically a 10 – turn potentiometer, coupled via an electromagnetic clutch to the chart drive shaft. It was used to sweep the control potential supplied to the potentiostat, the sweep was tied to chart speed with a maximum sensitivity of 100 mV/cm. by the use of optional potential divider unit, type DIDT, the sensitivity could be set at 25, 50 or 100 mV/cm. The experimental procedure which was based on the standard reference method for making potentiostatic polarization measurement, which was under the Jurisdiction of ASTM committee G – I on corrosion of Metals[110], and involved the following steps.

1- The specimen was mounted on the electrode holder and was further cleaned just, prior to immersion, by degreasing for 5 min in hot benzene, followed by acetone.

2- One liter of the sodium hydroxide solution at a given concentration was prepared from Analar grade base and distilled water. A 750 ml of the desired solution was transferred to the clean test cell [Fig. (2-8)].

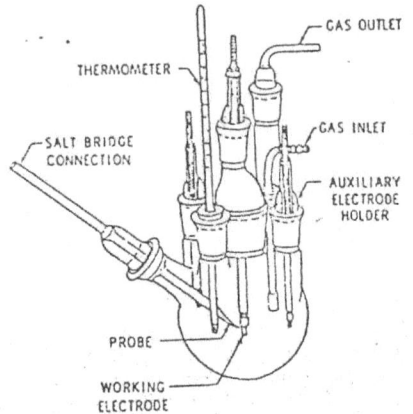

THERMOMETER

GAS OUTLET

SALT BRIDGE
CONNECTION

GAS INLET

AUXILIARY
ELECTRODE
HOLDER

PROBE

WORKING
ELECTRODE

Fig. (2-8) : *A schematic diagram of the polarization cell.*

<u>3-</u> The temperature of the solution was brought to the desired value by immersing the test cell in a controlled temperature water bath with a precision of $\pm$ 0.1°C, a temperature regulator called Temp – unit, type HAAKE – KT – 33, was used.

<u>4-</u> The Platinzed auxiliary electrodes, the Luggin bridge and other components were placed in the test cell by the usual procedures, the tip of the Luggin capillary was placed as close as physically possible to the surface of the working electrode in the corrosion cell. The Luggin bridge was filled with the test solution and temporarily close the center opening with a glass stopper.

70

5- The potential scan started 1 hour after the specimen immersion in the base solution, beginning at about (- 2 V) and proceeded through to (+ 2 V) versus the saturated calomel electrode (SCE). A potential against (*log* current density) was recorded by x – y recorder [Fig. (2-1) part A] at a potential scan rate of (0.3 V min^{-1}).

2-7 Chemicals And Materials

The aluminium alloys that have been studied in the present work were 2024, 5083, and 7075. The standard chemical compositions for these alloys are shown in appendix (page246).

Sodium hydroxide was used as the test solution with three concentration (0.1, 1x10^{-3}, and 1x10^{-5} mol.dm^{-3}) and sodium chloride was used to study the effect of chloride ions in solution with three concentration (1x10^{-3}, 1x10^{-2}, 0.1 mol.dm^{-3}).

Also sodium acetate and sodium chromate were used as organic and inorganic inhibitors with three concentration (0.05, 0.10, and 0.15 mol.dm^{-3}) and (5x10^{-3}, 1x10^{-2}, 5x10^{-2} mol.dm^{-3}) respectively, the molecular weight and a purity of materials were illustrated in Table (2-1).

Table (2-1) : *The molecular weight and purity of materials.*

Materials	M.wt (g.mol^{-1})	Purity (%)	Source
Sodium hydroxide	40.00	>99.5	FERAK
Sodium chloride	58.44	99.5	Fluka
Sodium acetate	82.03	98.0	BDH chemicals
Sodium chromate	161.99	>96.0	BDH chemicals

Chapter Three: Results and Discussion of
Polarization Behavior in Basic Media

3-1 Introduction

A convenient way to express the corrosion behaviour of a metal is by considering its potential – current density diagram which is generally known as the polarization curve, as presented in Figure (3-1). The curve (ABC) represents the cathodic region which continues up to a point corresponding to the corrosion potential (E_{corr}) of the metal, including the limiting concentration polarization section (AB) which is followed usually by a linear Tafel region[111]. The latter section (BC) relates the applied electrode potentials on the metal and the logarithm of current densities (log i_{corr}). This section involves in many cases the discharges of hydrogen ions forming hydrogen atoms on the electrode surface and a consequent formation of molecular hydrogen which librates at the electrode surface; this is usually an activation – energy controlled process. The section (CD) of the curve shows a remarkable increase of anodic dissolution current, corresponding to the active dissolution zone of the metal; the section (CD) presents the anodic Tafel region of polarization curve[111].

In the (DEG) section of the curve the anodic current sinks usually to a low value, in compliance with an exponential law, maintaining itself at a diminished value over a wide potential range; this is the passivity range region. The onset of passivity, corresponding to the formation of passive layer, occurs at a definite potential, called "Flade" potential.

The highest value of the anodic dissolution current before the metal (or the alloy) becomes passive (point D on the curve) represents the critical current density (i_{crit}). It makes an important change in the kinetics of the electrode reaction, the change from a high dissolution rate (active state)[111].

The critical current density may be considered as the limiting value at which occurs the anodic polarization necessary to reach the Flade potential. As the potential is increased, the current requirements increases again and corrosion resumes in the transpassive range along the section (GH) of the curve. The point (G) may correspond to the transpassive state; the potential and current density corresponding to this point and the transpassive potential and current density respectively[111]. The passivity region may have different other forms which may be referred to as :

i- The passivity region may disappear completely from the polarization curve particularly in acidic media.

ii- It may be composed of two or more passive regions, which are then termed: the first, the second ...... etc. Passive regions when such subdivision enters the polarization curve several critical and Flade potentials (E_f) will develop.

iii- Increasing the concentration of the electrolyte or of temperature in the corrosion medium may cause in general a shift of the whole polarization curve towards higher current densities.

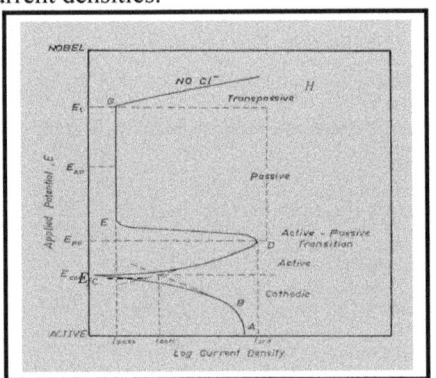

Fig. (3-1) : *A typical polarization diagram*
of a corroding system.

74

3-2 Results And Discussion Of Polarization Curves Of Pure Aluminium And Its Alloys In Basic Media

The various equilibria of Al – H_2O system has been collated by Pourbaix et. al in a potential versus pH diagram as shown in Figure (3-2).

This diagram indicates the theoretical circumstances in which aluminium should show corrosion (forming Al^{3+} at low pH value and AlO_2^- at high pH value),passivity due to hydragillite, i.e. $Al_2O_3.3H_2O$ (at near – neutral pH values) and immunity (at high negative potentials).

The nature of the oxide actually varies according to temperature, boehmite ($Al_2O_3.H_2O$) is the stable form.

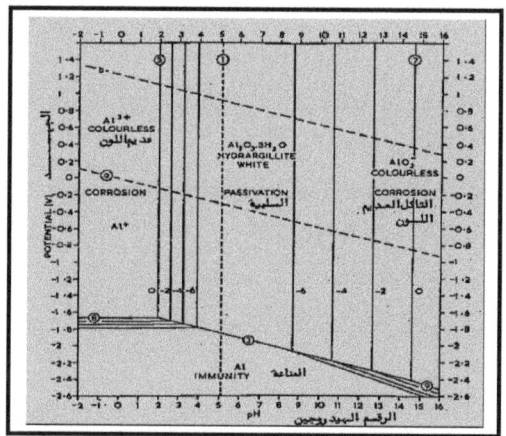

Fig. (3-2) : _Potential versus pH diagram for Al/H_2O system at 25°C[6]._

A typical polarization curve for pure Al metal was similar to the polarization curve of its alloys (2024, 5083, and 7075) at constant pH and temperature. The difference was showed in the corrosion potential and corrosion current density.

Figure (3-3) shows a typical potentiostatic curve for pure Al metal in 0.1 mol.dm^{-3} NaOH solution (pH = 13) at 298 K.

The section (ab) on the curve represents the diffusion and transport of cations (Al^{3+} and H_3O^+) from the bulk solution to the metal / solution interface as shown in the following reaction :

$$Al^{3+} + 3e \rightarrow Al$$

$$2H_2O \leftrightarrow H_3O^+ + OH^-$$

Before electron transfer can occur the oxygen must be transported to the metal/solution interface by diffusion and by natural and forced convection and reduction of oxygen can occur [(bc) section] as shown in the following reaction:

$$O_2 + 2H_2O + 4e \leftrightarrow 4OH^-$$

Anodic dissolution of pure Al and its alloys begin at point (c) and continue along (cd) forming AlO_2^- ion.

Along the section (def) the metal hydroxide is expected to be formed according to the following reactions :

$$Al + 3OH^- \rightarrow Al(OH)_3 + 3e$$

$$Al(OH)_3 + OH^- \rightarrow Al(OH)_4^-$$

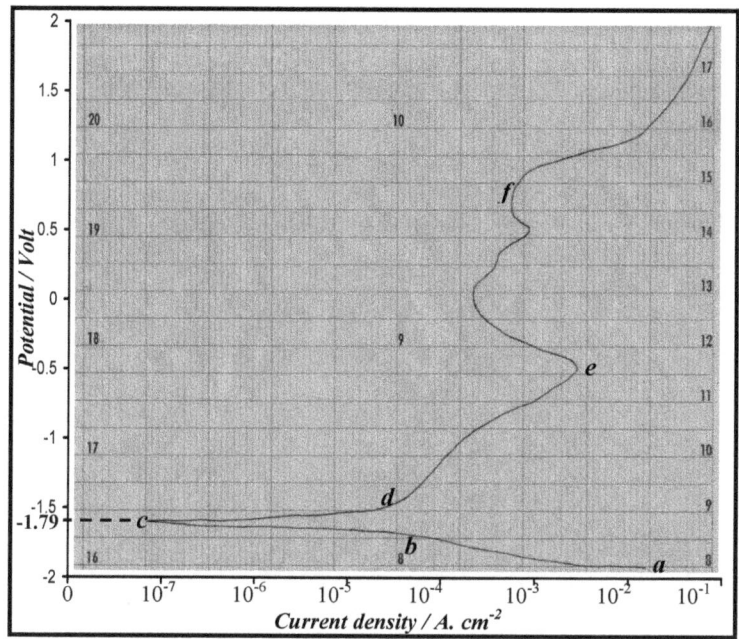

Fig. (3-3) : _The polarization curve of pure Al in 0.1 mol.dm^{-3}_
aerated NaOH solution (pH = 13) at 298K.

The polarization curve of pure Al in 1×10^{-3} mol.dm^{-3} NaOH solution (pH = 11) is shown in Fig. (3-4) which was generally similar to that obtained for Al alloys in the same media.

The cathodic Tafel region (abc), which can be attributed to the hydrogen evolution reaction by water discharge process in addition to the diffusion and transport of (Al^{3+}) to the metal / solution interface according to the reaction :

$$Al^{3+} + 3e \rightarrow Al$$
$$Al + 2H_2O \rightarrow AlOOH + 3/2\ H_2$$

In addition to the reduction of oxygen reaction:

$$O_2 + 2H_2O + 4e \leftrightarrow 4OH^-$$

77

The anodic Tafel region (cd) represents the anodic Tafel oxidation of Al to aluminium hydroxide. This result is a good agreement with the following main anodic reaction[112] :

$$Al + 2H_2O \rightarrow AlO_2^- + 4H^+ + 3e$$

The cathodic and anodic section in Fig. (3-4) are quite , sharp and clear as shown below.

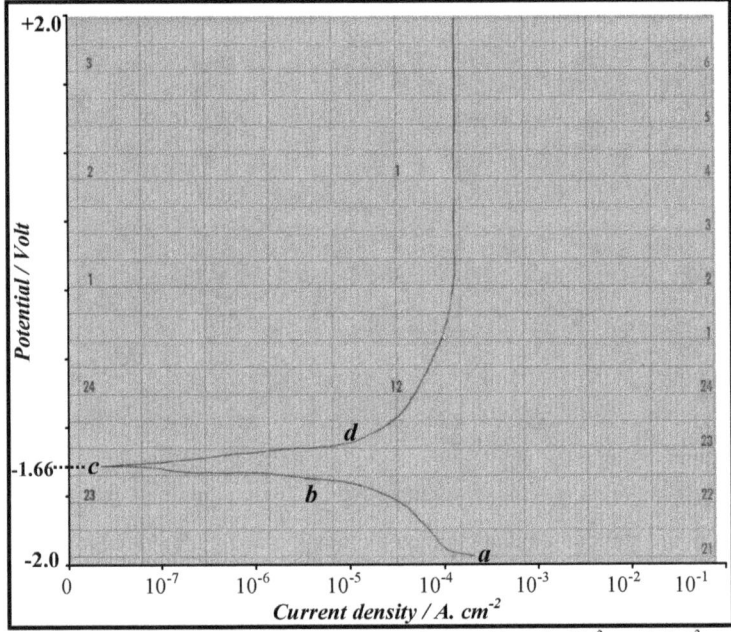

<u>**Fig. (3-4)**</u> : *The polarization curve of pure Al in $1x10^{-3}$ mol.dm^{-3} aerated NaOH solution (pH = 11) at 298K.*

In $1x10^{-5}$ mol.dm^{-3} NaOH solution (pH = 9), the polarization curve for pure Al and its alloys were quite different from that show in pH (13 and 11). A typical polarization curve in pH = 9 is shown in Fig. (3-5).

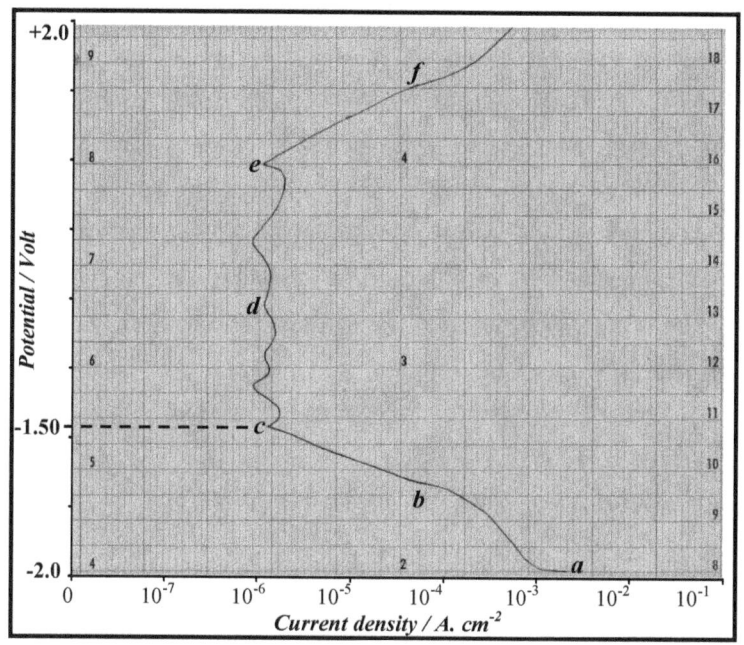

Fig. (3-5) : _The polarization curve of pure Al in $1x10^{-5}$ mol.dm^{-3} aerated NaOH solution (pH = 9) at 298K._

The section (abc) on the curve represents the diffusion and transport of cations and the discharge activation process of hydrogen ion with consequent evolution of hydrogen on the working electrode surface.

Along the section (cde), the metal hydroxide is expected to be formed. The hydroxide soon dissociates into metal oxide (Al_2O_3) on Al surface which behaves as a passive layer (protective film) according to the reaction :

$$Al^{3+} + 3H_2O \rightarrow Al_2O_3 + 6H^+$$

79

Generally, the solubility of the hydroxide film depends largely on the magnitude of the basicity or acidity. The nature and formation of film has been the subjects of numerous investigations[113-118].

It has been widely accepted that duplex type of film are formed. On immersion in water, oxide film can grow up to a thickness of $(10\ A^o)$.

The duplex film on aluminium [Fig. (3-6)] consist of a protective inner barrier oxide layer and a bulk (outer oxide) layer. The composition of amorphous bulk hydrated oxide film is dependent on the interaction corroding medium. The role of pH is important, as it affects the solubility as shown in Fig. (3-7).

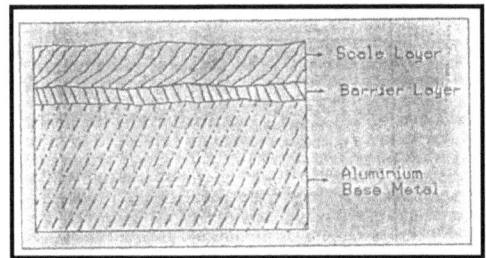

Fig. (3-6) *: Duplex film in aluminium*[53].

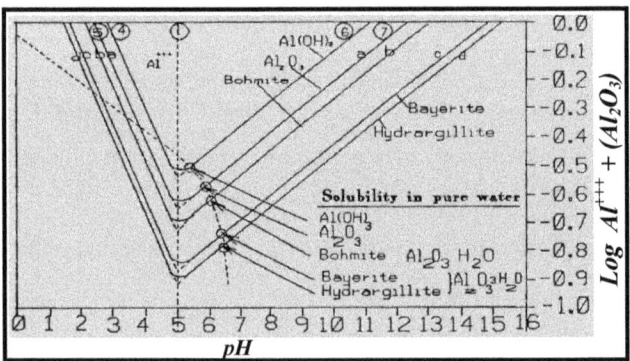

Fig. (3-7) *: Influence of pH on the solubility of Al_2O_3 and its hydrates at 25^oC*[53].

Also, it has been suggested that film formation of amorphous in three stages, the first being the formation of amorphous aluminium hydroxide $Al(OH)_3$, the second mono – hydrate orthorhombic crystalline boehmite γ- $Al_2O_3.3H_2O$ [or γ- $AlO(OH)$] and the third being the trihydrate monoclinic bayeite β- Al_2O_3 or hydragillite [1/2 $Al_2O_3.3H_2O$][119].

It has been suggested that the over all corrosion rate is attracted by the thickness and nature of the bulk film, which in most instances is a film of oxide layer which takes place are reflected by the change in the shapes of the anodic polarization curve.

The breakdown of passivity began at point (e) and continued along (ef) and this section represents the anodic region in Fig. (3-5).

The corrosion potential (E_{corr}) and the corrosion current densities (i_{corr}) have been obtained by the extrapolation of the Tafel lines to the points of intersections as shown in Fig. (3-1). Tables (3-1) to (3-4) present values of the corrosion potentials(Volt),corrosion current densities($A.cm^{-2}$), passivity potentials E_p (Volt), passivity current densities i_p ($A.cm^{-2}$), cathodic b_c and anodic b_a Tafel slopes (Volt. $decade^{-1}$), cathodic α_c and anodic α_a transfer coefficients and polarization resistances R_p ($\Omega.cm^{-2}$), for the polarization of pure aluminium and its three alloys (2024, 5083, and 7075) in three different concentrations of NaOH solution (pH = 13, 11, and 9) at four different temperatures (298, 303, 308, and 313)K.

Table (3-1) : *Corrosion parameters for the polarization of pure aluminium in aerated NaOH solution at four temperatures.*

pH	T (K)	Corrosion		b (V.decade^{-1})		α		$R_p/10^5$ Ω.cm^{-2}	$i_o/10^{-6}$ A.cm^{-2}	Passivity	
		$-E_{corr}$ (V)	$i_{corr}/10^{-5}$ A.cm^{-2}	$-b_c$	$+b_a$	$α_c$	$α_a$			$-E_p$ (V)	$i_p/10^{-6}$ A.cm^{-2}
13	298	1.79	0.39	0.035	0.070	1.689	0.844	4.58	0.055	-	-
	303	1.77	0.47	0.045	0.072	1.355	0.835	3.76	0.069	-	-
	308	1.75	0.55	0.046	0.074	1.328	0.825	3.18	0.083	-	-
	313	1.74	0.59	0.047	0.095	1.321	0.653	2.94	0.091	-	-
11	298	1.66	58.59	0.028	0.031	2.111	1.907	0.028	9.1678	-	-
	303	1.64	78.13	0.030	0.032	2.004	1.878	0.021	12.428	-	-
	308	1.63	97.66	0.034	0.035	3.675	1.745	0.017	15.611	-	-
	313	1.62	117.19	0.036	0.040	1.724	1.552	0.014	19.265	-	-
9	298	1.50	10.66	0.150	-	0.394	-	0.141	1.688	1.10	1.21
	303	1.42	11.03	0.186	-	0.323	-	0.129	2.023	0.95	1.29
	308	1.41	14.70	0.187	-	0.326	-	0.096	2.764	0.92	1.58
	313	1.40	15.44	0.189	-	0.328	-	0.091	2.963	0.90	1.75

Table (3-2) : *Corrosion parameters for the polarization of 2024 alloy in aerated NaOH solution at four temperatures.*

pH	T (K)	Corrosion		b (V.decade^{-1})		α		$R_p/10^6$ Ω.cm^{-2}	$i_o/10^{-7}$ A.cm^{-2}	Passivity	
		$-E_{corr}$ (V)	$i_{corr}/10^{-6}$ A.cm^{-2}	$-b_c$	$+b_a$	$α_c$	$α_a$			$-E_p$ (V)	$i_p/10^{-6}$ A.cm^{-2}
13	298	1.75	0.62	0.039	0.040	1.515	1.478	2.842	0.0903	-	-
	303	1.73	0.99	0.041	0.042	1.466	1.431	1.756	0.148	-	-
	308	1.70	1.72	0.046	0.053	1.328	1.222	0.992	0.267	-	-
	313	1.68	3.45	0.063	0.055	0.985	1.240	0.478	0.564	-	-
11	298	1.50	8.87	0.026	0.043	2.273	1.374	0.169	1.518	-	-
	303	1.46	17.24	0.029	0.048	2.072	1.252	0.085	3.070	-	-
	308	1.44	18.23	0.060	0.052	1.018	1.175	0.079	3.359	-	-
	313	1.42	19.70	0.063	0.054	0.985	1.149	0.072	3.745	-	-
9	298	1.39	0.95	0.135	-	0.437	-	1.460	0.175	0.78	1.13
	303	1.26	1.00	0.240	-	0.250	-	1.260	0.207	0.69	1.14
	308	1.23	1.08	0.448	-	0.136	-	1.138	0.233	0.67	1.15
	313	1.22	1.10	0.480	-	0.129	-	1.109	0.243	0.63	1.16

Table (3-3) : *Corrosion parameters for the polarization of 5083 alloy in aerated NaOH solution at four temperatures.*

pH	T (K)	Corrosion		b (V.decade^{-1})		α		$R_p/10^6$ Ω.cm^{-2}	$i_o/10^{-7}$ A.cm^{-2}	Passivity	
		$-E_{corr}$ (V)	$i_{corr}/10^{-6}$ A.cm^{-2}	$-b_c$	$+b_a$	$α_c$	$α_a$			$-E_p$ (V)	$i_p/10^{-6}$ A.cm^{-2}
13	298	1.80	0.60	0.031	0.034	1.907	1.738	3.01	0.0852	-	-
	303	1.79	1.92	0.037	0.036	1.624	1.669	1.94	0.133	-	-
	308	1.78	1.01	0.039	0.039	1.566	1.566	1.76	0.150	-	-
	313	1.77	1.15	0.043	0.053	1.444	1.171	1.54	0.175	-	-
11	298	1.66	8.74	0.034	0.031	1.738	1.907	0.19	1.351	-	-
	303	1.65	9.66	0.035	0.032	1.717	1.878	0.17	1.526	-	-
	308	1.64	11.49	0.036	0.034	1.697	1.797	0.14	1.855	-	-
	313	1.63	13.79	0.037	0.036	1.678	1.724	0.11	2.285	-	-
9	298	1.61	0.98	0.130	-	0.454	-	1.64	0.156	0.94	0.91
	303	1.57	1.03	0.137	-	0.438	-	1.52	0.171	0.91	1.05
	308	1.51	1.09	0.142	-	0.430	-	1.38	0.191	0.85	1.13
	313	1.42	1.12	0.152	-	0.408	-	1.26	0.212	0.79	1.14

Table (3-4) : *Corrosion parameters for the polarization of 7075 alloy in aerated NaOH solution at four temperatures.*

pH	T (K)	Corrosion		b (V.decade^{-1})		α		$R_p/10^6$ Ω.cm^{-2}	$i_o/10^{-7}$ A.cm^{-2}	Passivity	
		$-E_{corr}$ (V)	$i_{corr}/10^{-6}$ A.cm^{-2}	$-b_c$	$+b_a$	$α_c$	$α_a$			$-E_p$ (V)	$i_p/10^{-6}$ A.cm^{-2}
13	298	1.87	0.17	0.031	0.026	1.907	2.273	10.80	0..023	-	-
	303	1.86	0.28	0.034	0.036	1.768	1.669	6.72	0.038	-	-
	308	1.85	0.69	0.037	0.037	1.654	1.651	2.67	0.099	-	-
	313	1.84	0.90	0.047	0.043	1.321	1.444	2.04	0.131	-	-
11	298	1.75	6.23	0.031	0.041	1.907	1.442	0.28	0.913	-	-
	303	1.60	7.27	0.032	0.042	1.821	1.431	0.22	1.186	-	-
	308	1.56	8.65	0.035	0.043	1.745	1.421	0.18	1.474	-	-
	313	1.50	10.38	0.039	0.045	1.592	1.379	0.14	1.872	-	-
9	298	1.72	1.04	0.140	-	0.422	-	1.65	0.154	1.10	0.97
	303	1.71	1.11	0.156	-	0.385	-	1.54	0.169	0.99	1.07
	308	1.60	1.18	0.160	-	0.381	-	1.36	0.195	0.98	1.14
	313	1.50	1.21	0.171	-	0.363	-	1.03	0.262	0.97	1.51

3-3 Results And Discussion Of Corrosion Potentials (E_{corr})

The corrosion potential (E_{corr}) of a material in a certain medium at a constant temperature is a thermodynamic parameter which is a criterion for the extent of the corrosion feasibility under the equilibrium potential (in opposite sign) of the cell consisting of the working electrode and the auxiliary electrode when the rate of anodic dissolution of working electrode material becomes equal to the rate of the cathodic process that takes place on the same electrode surface.

When (E_{corr}) becomes more negative, the potential of the Galvanic cell becomes more positive and hence the Gibbs free energy change (ΔG) for the corrosion process becomes more negative. The corrosion reaction is then expected to be more spontaneous on pure thermodynamic ground.

When the measured value of (E_{corr}) becomes less negative, the potential of the corresponding Galvanic cell becomes less positive, hence the (ΔG) value for the corrosion process becomes less negative, and the process is thus less spontaneous.

It is thus shown that (E_{corr}) value is a measure for the extent of the feasibility of the corrosion reaction on purely thermodynamic basis. Values of (E_{corr}) for pure Al and its alloys in the three values of pH (13, 11, and 9) at 298K are plotted in Figs. (3-8) to (3-10).

To compare between corrosion resistance for pure aluminium and its alloys, the difference in the corrosion potentials are usually used for this purpose. The following order in the terms of increasing corrosion potential (decreasing corrosion) at constant pH and temperature :

Increasing (E_{corr})   *2024 alloy > pure Al $\simeq$ 5083 alloy > 7075 alloy*
decreasing the corrosion   *Al-Cu-Mg   99.99% Al   Al-Mg   Al-Zn-Mg*

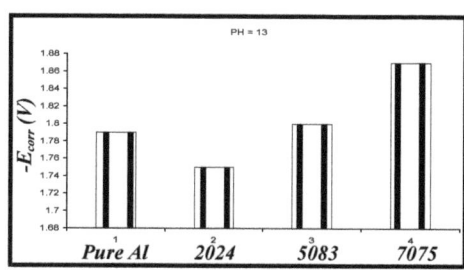

Fig. (3-8) *: Values of (E_{corr}) plotted for*
pure Al and its alloys in pH=13 at 298K.

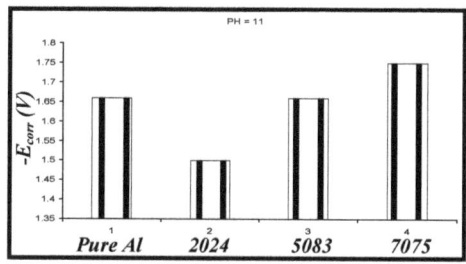

Fig. (3-9) *: Values of (E_{corr}) plotted for*
pure Al and its alloys in pH=11 at 298K.

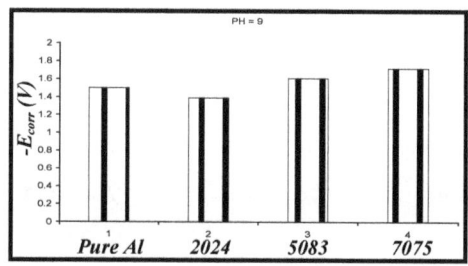

Fig. (3-10) *: Values of (E_{corr}) plotted for*
pure Al and its alloys in pH=9 at 298K.

The difference between corrosion resistance for pure Al and its alloys is due to the effect of alloying elements and the condition of the alloys, Al – Cu – Mg alloy (2024) has higher value of (E_{corr}) and so higher corrosion resistance than pure Al and other alloys.

The increasing (E_{corr}) value in 2024 alloy is attributed to the presence of copper and since most of the copper is present in the solid solution which increase the corrosion resistance of aluminium, and this behaviour can be attributed to the nature of passive film covering this alloy.

Also this effect attributed to the preferential dissolution of aluminium from the pit region leading to the formation of Cu – rich surface inside pits which shifts (E_{corr}) to more noble values.

Also the surface of this alloy displays black marks after the first few hours of corrosion, according to the reaction :

$$Cu \; + \; \tfrac{1}{2} \, O_2 \; \rightarrow \; CuO$$

E_{corr} values of Al – Mg alloy (5083) has close values to that of pure Al, which means that magnesium has very small effect on the resistance of aluminium, 5083 alloy has an (E_{corr}) values more active than that for pure aluminium.

This behaviour was attributed to the preferential dissolution of magnesium from the surface resulting in the exposure of new surface which is similar to that covering pure aluminium.

Al – Zn – Mg alloy (7075) has lower (E_{corr}) values (more negative) and so lower corrosion resistance than pure Al and other alloys (2024 and 5083).

This peculiar behaviour of the alloy 7075 can be attributed to the presence of the active $MgZn_2$ phase whose (E_{corr}) value is too active compared with the solid solution of this alloy, then by applying the desired potential,

MgZn$_2$ will begain to dissolve at high rate causing sudden increase in the current density[120].

After a short time, all MgZn$_2$ present in the surface will be completely consumed causing the alloy to behave in a normal way , i.e, by returning the current to passive region.

At a constant temperature, for same electrode (pure Al, 2024, 5083, or 7075), corrosion increases with increasing concentration of medium (NaOH solution) or increasing pH value toward basicity as shown in the Figs. (3-11) to (3-14), which show the effect of concentration of NaOH solution on the (E$_{corr}$) values, the following order indicates this behaviour :

Increasing (-E$_{corr}$)

[increasing corrosion   pH=13 > pH=11 > pH=9

in term pH value]

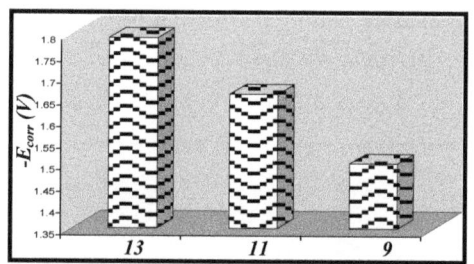

Fig. (3-11) : *Values of (E$_{corr}$) plotted against pH value for pure Al at 298K.*

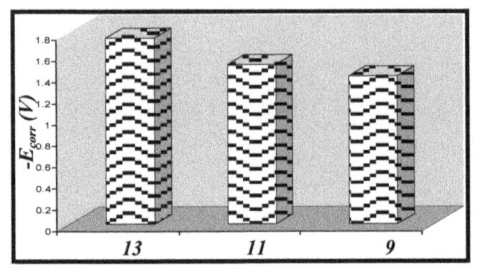

Fig. (3-12) : *Values of (E_{corr}) plotted against pH value for 2024 alloy at 298K.*

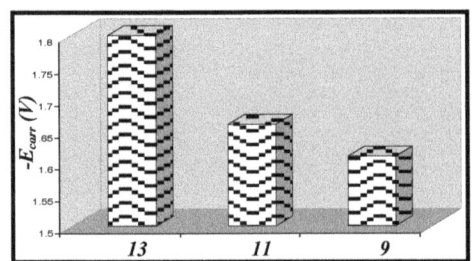

Fig. (3-13) : *Values of (E_{corr}) plotted against pH value for 5083 alloy at 298K.*

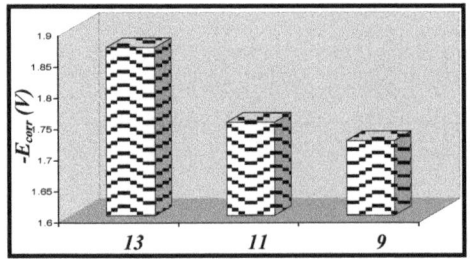

Fig. (3-14) : *Values of (E_{corr}) plotted against pH value for 7075 alloy at 298K.*

3-4 Results And Discussion Of Current Densities (i_{corr})

The corrosion current density (i_{corr}) is a kinetic parameter and represents the rate of corrosion under specified equilibrium condition.

Any factor that enhances the value of (i_{corr}) results in an enhanced value of the corrosion rate on pure kinetic ground. Generally, the data in the Tables (3-1) to (3-4) indicates that (i_{corr}) for pure Al higher than (i_{corr}) of alloys in three experimental values of pH and the (i_{corr}) value takes the following order, also the (i_{corr}) increases with increasing temperature as shown in Tables (3-1) to (3-4);

At pH=13 and 11 i_{corr} *(A.cm⁻²)*   *2024 > 5083 > 7075*

While

At pH=9 i_{corr} *(A.cm⁻²)*   *7075 > 5083 > 2024*

At constant temperature, for same electrode (pure Al, 2024, 5083, or 7075), (i_{corr}) values vary with increasing concentration of medium (NaOH solution) or increasing pH value, this variation are plotted in Figs. (3-15) to (3-17) which indicates that :

i_{corr} at constant temp.

for same electrode   *11 > 9 > 13*

in term pH values

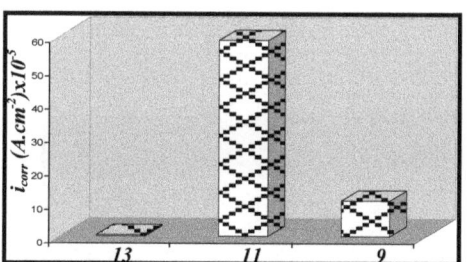

Fig. (3-15) : *Values of (i_{corr}) plotted against pH value for pure aluminium at 298K.*

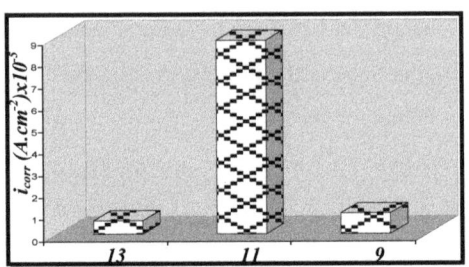

Fig. (3-16) *: Values of (i_corr) plotted against*
pH value for 2024 alloy at 298K.

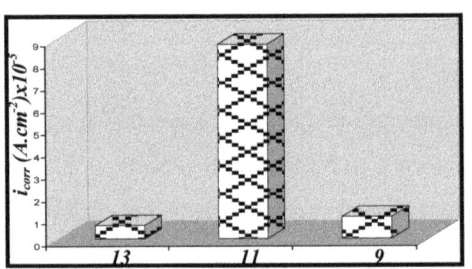

Fig. (3-17) *: Values of (i_corr) plotted against*
pH value for 5083 alloy at 298K.

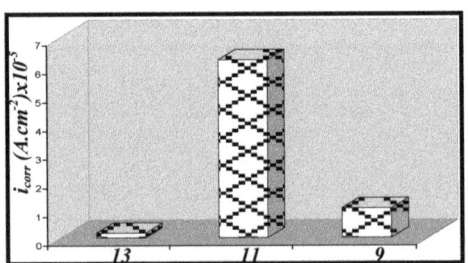

Fig. (3-18) *: Values of (i_corr) plotted against*
pH value for 7075 alloy at 298K.

3-5 Passive Potentials (E_p) And Passive Current Densities (i_p)

Passivity is an unusual phenomenon observed during the corrosion of certain metals and alloys, it can be defined as a loss of chemical reactivity under certain environmental conditions[121].

According to the Fig. (3-2), in the experimental condition, the passivity appears only in pH = 9 and the data are indicated in Tables (3-1) to (3-4). The sequence of the increasing (E_p) values for different working electrodes in three pH values was as :

$$-E_p \; (V) \qquad 2024 > 5083 > pure \; Al > 7075$$

The passive potential of a passive film on a metal or alloy depends upon the nature of the metal and alloy, it becomes less positive depending on the stability of the existing oxide film.

The presence of certain anions destroys the passivity and results in localized corrosion[122, 123].

While the sequence of (i_p) values for the different working electrodes in three pH values was as :

$$i_p \; (A.cm^{-2}) \qquad pure \; Al > 2024 > 7075 > 5083$$

The decreasing passive current density (i_p) for the electrodes may be connected with the increasing stability of the oxide film, while the increasing in (i_p) implies a decrease in the stability of the oxide film which tends to dissociate at the end close to the transpassive potential[124].

3-6 Tafel Slope (b) and Transfer Coefficient (α)

From deep analysis of the cathodic, anodic and passive regions of the polarization curves which have been obtained for pure Al and its alloys in three values of pH at four temperatures, it was possible to derive data concerning :

a- The cathodic (b_c) and anodic (b_a) Tafel slopes.

b- The cathodic (α_c) and anodic (α_a) transfer coefficients.

Values of α have been calculated from the corresponding values of the Tafel slope (b) using the relation[125, 35]:

$$\alpha_c = \frac{2.303RT}{b_c F} \qquad\qquad \text{.........(3-1)}$$

$$\alpha_a = \frac{2.303RT}{b_a F} \qquad\qquad \text{.........(3-2)}$$

Values of Tafel slopes (b_c or b_a) for both cathodic and anodic reactions were generally close to (0.120 V.decade^{-1}) and the corresponding values of the transfer coefficients (α_c and α_a) were close to (0.5).

The main exception to this result was the relatively some higher or lower values of the Tafel slopes (b_c or b_a) or of the transfer coefficients (α_c and α_a) for certain specimens in basic medium of NaOH solution with three values of pH.

A value of the cathodic transfer coefficient α_c of $\approx$ 0.5, or of the cathodic Tafel slope of (- 0.120 V.decade^{-1}), may be diagnostic of a proton discharge – chemical desorption mechanism in which the proton discharge is the rate – determining step (r.d.s).

The two basic reactions paths for hydrogen evolution reaction are:

$$H_3O^+_{\text{(bulk solution)}} \xrightarrow{\text{Diffusion}} H_3O^+_{\text{(metal/solution interface)}}$$

which is followed by the discharge step (D):

$$H_3O^+ + M + e^- \xrightarrow{D} M{-}H + H_2O$$

where M is the metal electrode and $M{-}H$ represents a hydrogen atom which is adsorbed on the metal surface. The discharge (D) step is usually followed by a chemical desorption (C$-$D) step as:

$$M{-}H + M{-}H \xrightarrow{C{-}D} 2M + H_2$$

In which two adjacent adsorbed hydrogen atoms unite together to form one molecule of gaseous hydrogen. If the chemical desorption is the rate – determining step, the rate would be independent of the overpotential since no charge transfer occurs in such a step and the rate becomes directly proportional to the concentration or the coverage (θ) of adsorbed hydrogen atoms[126,127]. On the other hand, if the discharge process is followed by a rate – determining step involving chemical desorption, the expected value of α should be (2.0).

In some cases, the previous two steps (D) and (C–D) may unite together to form one electrochemical desorption (E–D) step as:

$$M\!-\!H \ + \ H_3O+ \ M_{(electrode)} \xrightarrow{\ E\!-\!D\ } 2M + \ H_2 \ + \ H_2O$$

When electrochemical desorption becomes the rate – determining step for hydrogen evolution reaction on the cathode, the expected value of α will be (1.5). Two mechanisms have been proposed for the formation of precursor passive film on the materials. The first is the precipitation – oxidation mechanism and the second is the solid state mechanism, the latter mechanism would not be mass transfer affected, but would account for the formation of the precursor film[126,127].

The results of the Tables (3-1) to (3-4) indicate that the variation of the Tafel slopes and of the corresponding transfer coefficients could be interpreted in terms of the variation in the nature of the rate – determining step from charge transfer process to either chemical – desorption or to electrochemical desorption. The results of b and α for pure Al and its alloys can be summarized in the following order:

In pH=13 and 11	b_c *and* b_a << *0.120*
	α_c *and* α_a >>*0.5*
But in pH=9	b_c > *0.120*
	α_c <*0.5*

3-7 Polarization Resistance R_p

The polarization resistance, R_p, of according electrode is defined as the slope of a potential (E) – current density (i) plot of the corrosion potential (E_{corr}) as :

$$R_p = \left(\frac{\partial \eta}{\partial i}\right)_{T,C} \quad at \; \eta \rightarrow 0 \quad \ldots\ldots\ldots\ldots(3\text{-}3)$$

Where $\eta = E - E_{corr}$, is the extent of polarization of the corrosion potential and (i) is the current density (c.d) corresponding to a particular value of (η). From the polarization resistance, R_p the corrosion current density (c.d) i_{corr} can be calculated as :

$$i_{corr} = \frac{\beta}{R_p} \quad \ldots\ldots\ldots(3\text{-}4)$$

Where β is a combination of the anodic and cathodic Tafel slopes (b_a, b_c) as :

$$\beta = \frac{b_a b_c}{2.303(b_a + b_c)} \quad \ldots\ldots\ldots(3\text{-}5)$$

For the general case, by inserting equation (3-4) into equation (3-5) one obtains the so – called the Stern – Geary equation[128]:

$$R_p = \frac{b_a b_c}{2.303(b_a + b_c) i_{corr}} \quad \ldots\ldots\ldots(3\text{-}6)$$

The results of Tables (3-1) to (3-4) show that the polarization resistance for the corrosion of pure Al and its alloys in basic media. The results of R_p can be summarized as follow :

1- R_p for pure aluminium is greater than its values of alloys (2024, 5083, and 7075) in three values of pH (13, 11, and 9).

2- The sequence of R_p for aluminium alloys in the three basic medium of NaOH solution was in the order as follow :

$$R_p \quad \textit{7075 alloy} > \textit{5083 alloy} > \textit{2024 alloy}$$

3- To the same electrode at constant temperature R_p varies with the concentration of NaOH solution as follow :

$$R_p \quad pH=13 > pH=9 > pH=11$$

4- Generally, to the same electrode at constant pH, decreases R_p with increasing the temperature.

Chapter Four: Results and Discussion of Effect Chloride ions
4-1 Introduction

A compact passive film on a metal surface, which usually results from marked polarization of a metal, substantial resists corrosion of the metal in a given environment[125].

It is generally agreed that anodic passivation is caused by the usually oxide, that to a great extent reduces the metal dissolution[111]. Such a passivity may be represented by (CD) region of the polarization curve as shown in Fig. (4-1).

When the potential of passivated anode is raised, the simplest transpassivity which is represented by (DE) in Fig. (4-1) occurs where the passivity film is anodically oxidized to a soluble oxide. Local breakdown of passivity at a less positive potential than that required for transpassivity is often induced by the presence of halide ions, which chloride ion is evidently the most important in practice, and the resulting local pitting frequently becomes rapid[111]. The breakdown potential of a passive film on a metal or alloy depends upon the metal or the alloy composition, it becomes less positive as chloride ion concentration in the medium increases, and it may often be brought below the passivation potential, so that even partial passivity is no longer possible. The presence of certain anions particularly chloride, destroys passivity and lead to localized corrosion, which in many cases takes the form of pitting[111].

A better and more comprehensive explanation of the breakdown of passivity phenomenon may be treated as :

1- The aggressive anion, adsorbed on the oxide film surface, begins to enter and penetrates into film under the influence of the electrostatic field across the film/solution interface, when the field reaches a critical value corresponding to

97

the breakdown potential. It is usually found that (Cl^-) ions are more aggressive than other halide ions[111].

The initial entry of anions occurs at regions of the film corresponding to grain boundary, which carry film of less perfection, and in which more easily entered than into the more order region. They may travel completely through the passivating film, or the metal cations may diffuse outwards to meet them. A "contaminated oxide" film thus produced with a greatly increased ion conductivity[111].

2- The anions and cations passing through the "contaminated oxide" film have in general different nobilities.

The initiation of a second breakdown at "weak" film at or near a grain – boundary or other imperfection will therefore usually occur at a distance from the first, where the space charge of the first breakdown does not unfavorably influence further anion entry[111].

Discrete pits, as usually found, are thus initiated. The initiation of pitting will therefore be completely controlled by film breakdown in the active region. Weak points in the film would thus be the most active sites for pitting, and these may or may not be related to the substrate structure.

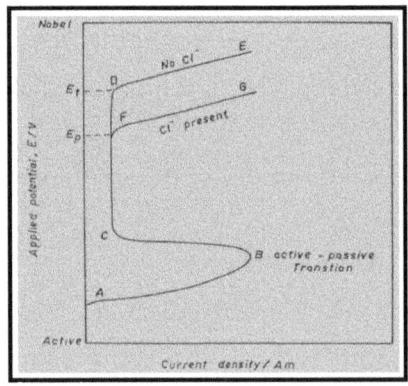

Fig. (4-1) : Schematic anodic Polarization curve showing the anodic passivity and passivity breakdown.

98

4-2 Theories And Mechanism Of Corrosion In Aluminium

The exceptional corrosion resistance of aluminium in many environments is due to its protective oxide film which is relatively inert chemically and so provide the passive behaviour of aluminium. However in aggressive environments, particularly in the presence of halide ions, aluminium suffers from localized corrosion by local breakdown of the passive film. Various theories are postulated to explain the mechanism of passivity breakdown in halide solutions but most common and accepted ones are as:

1- Penetration theory.

2- Flaws and crack/heal theory.

3- Localized acidification theory.

4- Complex formation theory.

5- Other models.

Penetration theory proposed that halide ions penetrate the passive film at certain points causing the breakdown of the passive layer at these points. Such penetration is proposed to take place by various models which can be summarized as follows:

a- Ionic migration under electric field

This model proposed a migration of the aggressive anions through the passive film lattice under the influence of electrostatic field or by anion exchange[129-137].

Hoar et al [111] proposed that the aggressive anions are adsorbed on the passive film after which they penetrate it without ion exchange under the electrostatic field corresponding to the breakdown potential. This process results in great increase in the ionic conductivity of the passive film at these points and the consequence is either the anions penetrate the passive film completely and attack bare metal directly or the cations migrate from the metal – oxide interface through the contaminated film to meet the aggressive anions. Finally, the occurrence of either of these two processes lead to dissolve the metal ions at certain points i.e pit nucleation.

_b-_ Gradual penetration, field independent

This model proposed that pits initiate as a result of gradual penetration of the chloride ions through the passive film[133-139]. Nguyen and Foley [129] suggested that the gradual penetration may occur by diffusion of halide ions through the film (lattice diffusion) or by migration through pores and fissures in the film, then attacking bare metal directly.

This model is confirmed by the work of Beck et al [138] and Heine et al [139] who used the impedance measurements and also by Painot and Augustynski [134,135], Augustynski [133], and Koudelkova and Augustynski [136] who used X – ray photoelectron spectroscopy.

All of these investigators detected the presence of chloride ions in the passive film lattice after an exposure to chloride solution.

_c-_ Anion exchange

This model suggest that pit initiate at the points where the oxygen adsorbed on the surface is displaced by chloride ions[140-145]. The displacement reaction occurs at the spots where the metal – oxygen bond is the weakest[146] and where the aggressive anion concentration exceeds a critical value, noticeable greater than its bulk concentration. According to this model Smialowska [140] and Kolotyrkin [141] defined the breakdown potential as the minimum electrode potential value at which the aggressive anions becomes capable to produce a reversible surface.

_d-_ Mechanical breakdown model

This model was proposed by Hoar[132] who suggested that the adsorbed anions push one another causing slipping and cracking of the passive film to which they are strongly adsorbed. Such cracks result in the exposure of bare metal which is then attacked by the aggressive anions directly leading to pit nucleation.

The above mode have the following drawbacks :

1- Metal and chloride ions can pass through the passive film only under the influence of high electric field while in the absence of this field, the transport through the passive film can be neglected at room temperature[147-150].

2- The displacement of oxygen ions by chloride ion seems to be unlikely because the oxygen ions must migrate against this high field [151].

3- The model which suggests that chloride ions are small enough to penetrate the passive film through pores and fissures is not accepted since large anions such as ClO_4^-, and SO_4^{2-} can cause pitting also and can not be expected to pass through the passive film.

The pitting of aluminium in chloride – containing waters follows a similar mechanism to that of steels [Fig. (4-2)], and again the characteristic feature of the process is the formation of acid within the occluded cell[152].

The passivating film of Al_2O_3 surrounding the pit acts as the cathode, but its effective in reducing dissolved oxygen is significantly enhanced if copper is either deposited on the surface or enters the lattice of the Al_2O_3, and it is well known that the pitting of aluminium occurs rapidly when the water contains a trace of copper ions. Similar considerations apply to intermetallic phase such as $FeAl_3$ and $CuAl_2$, which can increase the kinetics of oxygen reduction.

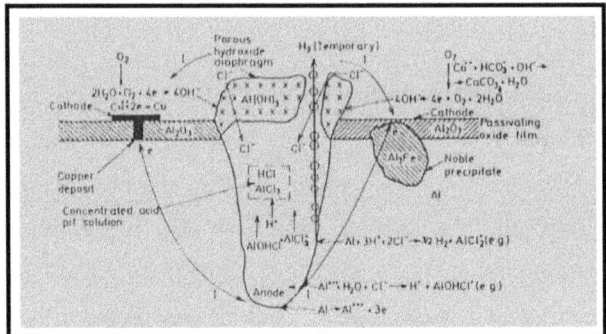

Fig. (4-2) : *Electrochemical mechanism of*

pit growth on aluminium.

4-3 Results And Discussion Of Polarization Curves

A typical polarization curve of pure Al in 0.1 mol.dm^{-3} NaOH (pH = 13) in the presence of 1x10^{-3} mol.dm^{-3} NaCl at 298K shown in Fig. (4-3) which was similar to that of Al alloys in the presence of three different concentrations of NaCl (1x10^{-3}, 1x10^{-2}, and 0.1 mol.dm^{-3}) at four different temperatures.

The cathodic (abc) and anodic (cd) sections are quite sharp and clear. The cathodic section involves the transport of cations (Al^{3+} and H$_3$O$^+$) from the bulk solution to the metal/solution interface and the discharge of hydrogen on the working electrode surface in addition to the adsorption and absorption of chloride ions on the working electrode.

Generally, there are no great difference in the behaviour of pure Al and its alloys in the absence and presence of chloride ion in pH=13.

Fig. (4-4) shows the addition of NaCl to 1x10^{-3} mol.dm^{-3} NaOH solution (pH = 11) for pure Al which was similar to that obtained for Al alloys at constant NaCl concentration.

At low concentration of NaCl (1×10^{-3} mol.dm^{-3}), the polarization curves consisted mainly of cathodic and anodic Tafel regions which were clear and smooth.

In the presence of (1×10^{-2} mol.dm^{-3}) NaCl, the curve consists of clear cathodic (bc) and anodic (cd) regions. The pure metal (and alloys) passes thereafter into the critical state (at point d) from which it is converted into a stable passive state at (e).

After to the passive layer which is formed at (e) is converted into a less stable passivity state along (fg). At (g), the less stable passive layer breaks down and then continues breaking along (gh).

The three stages which appear in this case are summarized in the following :

1- Cathodic reaction, was represented in (abc) section and include the following reactions :

$$Al^{3+} + 3e \rightarrow Al$$
$$O_2 + 2H_2O + 4e \leftrightarrow 4OH^-$$

2- Absorption and adsorption of chloride ion on the electrode surface, which was weak layer in the basic media and represented in (cde) section with the following reactions :

$$Al^{3+} + 2OH^- + Cl^- \rightarrow Al(OH)_2Cl$$
$$Or \quad Al^{3+} + 2OH^- + 2Cl^- \rightarrow Al(OH)_2Cl_2$$

3- Anodic reaction, destroy the weak passivity layer and dissolves the bare surface represented in the (efg) section with the following reaction :

$$Al + 2H_2O \rightarrow AlO_2^- + 4H^+ + 3e$$

At high concentration of chloride ion (0.1 mol.dm^{-3} NaCl), the polarization curve was generally similar to that observed in Fig. (4-3) for (0.1 NaOH + 1×10^{-3} NaCl) system as shown in Fig. (4-4).

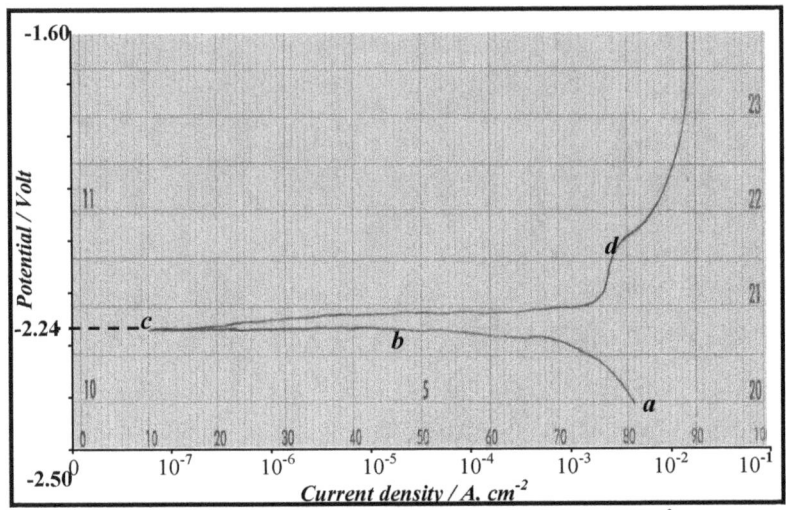

Fig. (4-3) : *The polarization curve of pure Al in 0.1 mol.dm^{-3} NaOH (pH=13) in presence of 1x10^{-3} mol.dm^{-3} NaCl at 298K.*

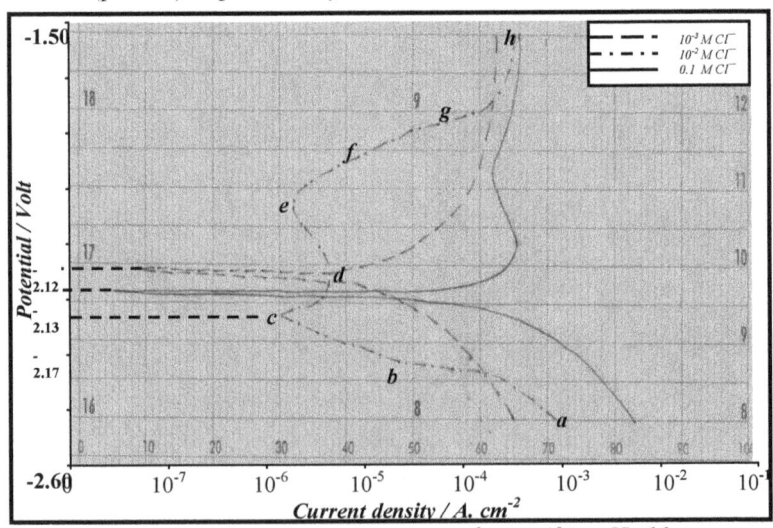

Fig. (4-4) : *The polarization curve of pure Al in pH=11 in presence of NaCl with three different concentration (1x10^{-3}, 1x10^{-2}, and 0.1 mol.dm^{-3}) at 298K.*

According to the diagram (3-2), the polarization behaviour of aluminium and its alloy have a protective film (passivation $Al_2O_3.3H_2O$) at neutral or near-neutral medium (such as pH=9). But the presence of chloride ions penetrates this passive layer and destroyes it.

Complex formation theory proposed that chemical reaction takes place between the halide ions and the passive film which leads to the formation of transitional complexes such as $Al_2Cl_7^-$, $Al(OH)_2Cl_2^-$, and $AlCl_4^-$ and finally attack bare metal directly.

These complexes are proposed to form at the points where sufficient halide ions jointly adsorbed to form halides island [153,154]. Hoar and Jacob (1967)[154] formulated this theory as follows:

Three or four halides ions are adsorbed on the passive film around the lattice cation (Al^{3+}), then they form a transitional complex with this cation and separate from the oxide ions in the lattice, but under the influence of the anodic field another cation comes up through the passive film to replace the dissolved one, and once the cation reachs the film – solution interface, it will find many halide ions with which it forms another transitional complex.

So, once the process begins it will repeat itself at an accelerated rate due to the increased electrostatic field on this thinned point. More recently Nguyen and Foley (1979)[129] suggests that pitting initiation process can be formulated by four consecutive steps according to this theory:

<u>1-</u> The halide ions are adsorbed on the passive film.

<u>2-</u> The chemical reaction between the adsorbed ions and (Al^{3+}) in the passive film lattice.

$$Al^{3+} \text{ (in } Al_2O_3.nH_2O \text{ lattice)} + Cl^- \rightarrow Al(OH)_2Cl$$

or

$$Al^{3+} \text{ (in } Al_2O_3.nH_2O \text{ lattice)} + 2Cl^- \rightarrow Al(OH)_2Cl_2^-$$

105

3- Thinning of the passive film by dissolution.

4- Direct interaction between the exposed metal and the halide ions to form transitional complexes which rapidly undergo hydrolysis :

$$Al^{3+} + 4 Cl^- \rightarrow AlCl_4^-$$

$$AlCl_4^- + 2H_2O \rightarrow Al(OH)_2Cl + 2H^+ + 3Cl^-$$

Figure (4-5) shows the polarization curves of pure Al and its alloys in pH=9 (1×10^{-5} mol.dm^{-3} NaOH) with the presence of (1×10^{-3} mol.dm^{-3} NaCl). The section (abc) on the curves represents the cathodic Tafel region, along the section (cde), the metal hydroxide is expected to formed.

The hydroxide soon dissociates into metal oxide (Al$_2$O$_3$) on aluminium surface which behaves as passive layer.

The breakdown of passivity began at point (e) and continued along (ef), and this section represents the anodic Tafel region.

The presence of 1×10^{-2} mol.dm^{-3} NaCl in pH =9, remains 7075 alloy only keeping passive layer while pure Al and 2024, 5083 alloys behave another behaviour as shown in Fig. (4-6). The section (abc) on the curves represents the wide cathodic Tafel region. The presence of chloride ions with this concentration break down any passive layer may be produced.

Anodic dissolution of pure Al begins at point (c) and continues along (de) according to the following reaction :

$Al + 3OH^- \rightarrow Al(OH)_3 + 3e$

$Al(OH)_3 + OH^- \rightarrow Al(OH)_4^-$

$Al(OH)_4^- + Cl^- \rightarrow Al(OH)_3Cl^- + OH^-$

$Al(OH)_3Cl^- + Cl^- \rightarrow Al(OH)_2Cl_2^- + OH^-$ $\left.\begin{array}{l}\\ \\ \end{array}\right\}$ or $Al(OH)_4^- + 4Cl^-$

$Al(OH)_2Cl_2^- + Cl^- \rightarrow Al(OH)Cl_3^- + OH^-$ $\qquad\qquad\downarrow$

$Al(OH) Cl_3^- + Cl^- \rightarrow AlCl_4^- + OH^-$ $\qquad\qquad AlCl_4^- + OH^-$

The stability of oxide layer extends over a wide range of potentials, but it is significantly lowered by the presence of chloride ions in the solution and the extent decreases with increasing the concentration of the chloride ions to $(0.1 \text{ mol.dm}^{-3})$, as shown in Fig. (4-7), where the passive region disappears completely.

The presence of chloride ions in the corrosion medium does not alter the nature of reaction taking place at the electrode. The chloride ions penetrate into the protective oxide film through pores.

In fact, as the chloride ion concentration increases, the surface coverage by such ions increases. The attack initially occurs along the grain boundaries and then spreads out into the grain interior.

As the concentration of the chloride ions increases, the rate of film breakdown approaches the rate of film formation. The passive film in such cases would be expected to be extremely thin and the passive region of the polarization curve will rather be small; the passive film does not then remain permanently intact. Instead, it continuously breaks down completely.

The results of E_{corr}, i_{corr}, b_c, b_a, α_c, α_a, E_p, i_p, i_o, and R_p which have been derived from the polarization curves for pure aluminium and its alloys in the NaOH solution with three pH values in the presence of chloride ion with three concentrations at four temperatures have been given in Tables (4-1) to (4-12).

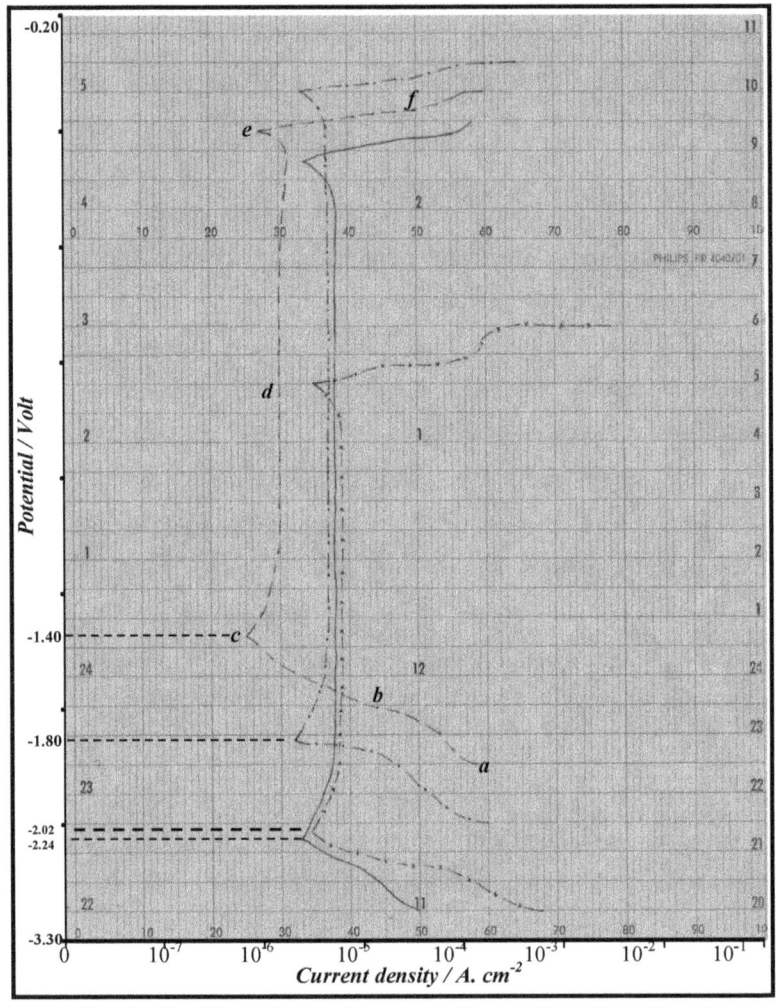

Fig. (4-5) : *The polarization curve of pure Al and its alloys in*
1x10⁻⁵ mol.dm⁻³NaOH (pH=9) in presence of
(1x10⁻³ mol.dm⁻³)NaCl at 298K.

——— *pure aluminium,* − − − *2024 alloy*

—ᐧ—ᐧ— *5083 alloy ,* —×—×— *7075 alloy*

108

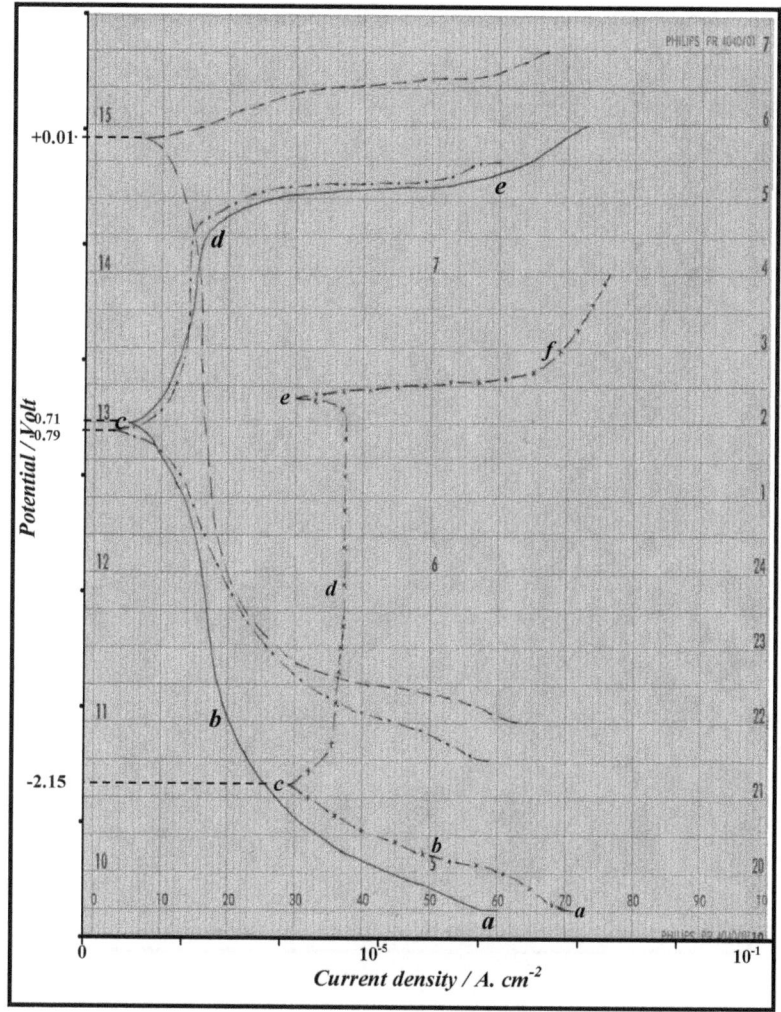

Fig. (4-6) : *The polarization curve of pure Al and its alloys in*
$1x10^{-5}$ mol.dm^{-3} NaOH (pH=9) in presence of
($1x10^{-2}$ mol.dm^{-3}) NaCl at 298K.
·—·— pure aluminium, — — — 2024 alloy
———— 5083 alloy , —×—×— 7075 alloy

109

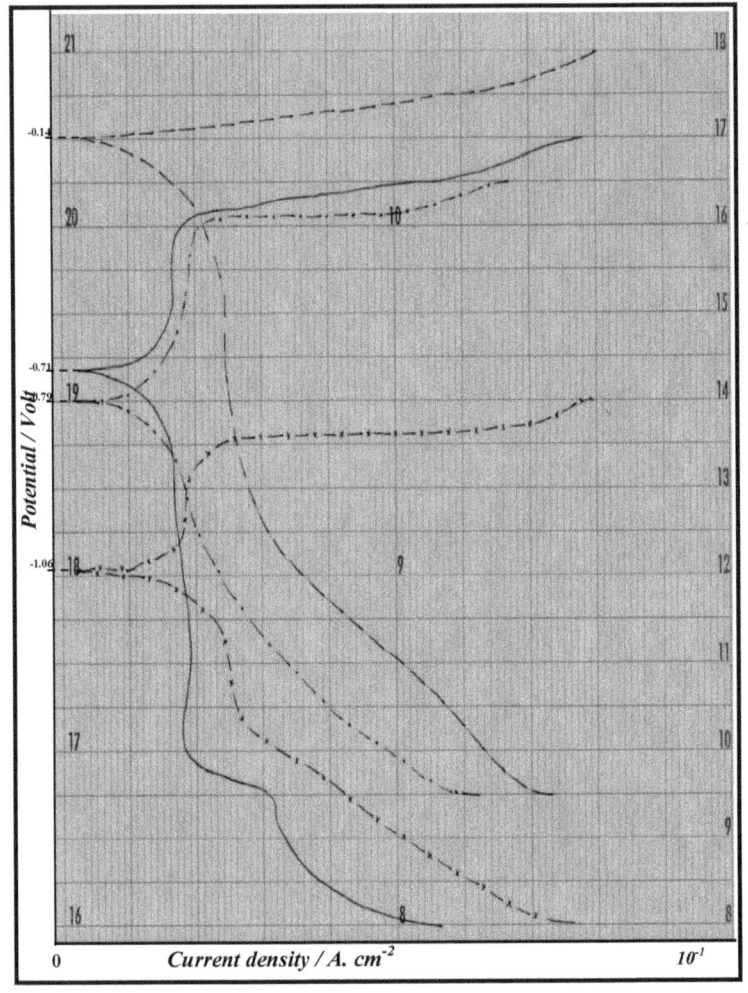

Fig. (4-7) : *The polarization curve of pure Al and its alloys in*
1x10⁻⁵ mol.dm⁻³NaOH (pH=9) in presence of
(0.1 mol.dm⁻³)NaCl at 298K.

───── *pure aluminium,* ─ ─ ─ *2024 alloy*
─·─·─ *5083 alloy ,* ─×─×─ *7075 alloy*

110

Table (4-1) : *Corrosion parameters for the polarization of pure aluminium in aerated NaOH solution pH=13 with presence of NaCl at four temperatures.*

Conc. Of NaCl mol.dm^{-3}	T (K)	Corrosion		b (V.decade^{-1})		α		$R_p/10^3$ (Ω.cm^{-2})	$i_o/10^{-5}$ (A.cm^{-2})
		$-E_{corr}$ (V)	$i_{corr}/10^{-3}$ (A.cm^{-2})	$-b_c$	$+b_a$	α_c	α_a		
$1x10^{-3}$	298	2.24	0.313	0.041	0.053	1.442	1.116	7.16	0.358
	303	2.23	0.515	0.048	0.067	1.252	0.897	4.33	0.603
	308	2.21	0.588	0.055	0.081	1.111	0.754	3.76	0.706
	313	2.20	0.662	0.067	0.094	0.927	0.661	3.32	0.812
$1x10^{-2}$	298	2.23	0.625	0.130	0.088	0.455	0.672	3.57	0.719
	303	2.21	0.699	0.141	0.111	0.426	0.542	3.16	0.826
	308	2.18	1.397	0.158	0.136	0.387	0.449	1.56	1.701
	313	2.17	1.838	0.162	0.145	0.383	0.428	1.18	2.285
0.1	298	2.19	0.651	0.065	0.066	0.909	0.895	3.36	0.764
	303	2.18	0.911	0.071	0.070	0.847	0.859	2.39	1.092
	308	2.17	1.253	0.083	0.081	0.736	0.754	1.73	1.534
	313	2.16	1.355	0.111	0.100	0.559	0.621	1.59	1.696

Table (4-2) : *Corrosion parameters for the polarization of 2024 alloy in aerated NaOH solution pH=13 with presence of NaCl at four temperatures.*

Conc. Of NaCl mol.dm^{-3}	T (K)	Corrosion		b (V.decade^{-1})		α		$R_p/10^3$ (Ω.cm^{-2})	$i_o/10^{-5}$ (A.cm^{-2})
		$-E_{corr}$ (V)	$i_{corr}/10^{-3}$ (A.cm^{-2})	$-b_c$	$+b_a$	α_c	α_a		
$1x10^{-3}$	298	1.91	0.123	0.044	0.040	1.344	1.478	15.53	0.165
	303	1.79	0.148	0.051	0.050	1.179	1.202	12.09	0.216
	308	1.69	0.197	0.067	0.068	0.912	0.898	8.58	0.309
	313	1.63	0.291	0.136	0.079	0.457	0.786	5.60	0.482
$1x10^{-2}$	298	1.79	0.201	0.081	0.080	0.729	0.739	11.85	0.217
	303	1.74	0.211	0.092	0.090	0.653	0.668	8.25	0.316
	308	1.68	0.394	0.107	0.103	0.571	0.593	4.26	0.623
	313	1.62	0.591	0.120	0.110	0.517	0.565	2.74	0.984
0.1	298	1.78	0.542	0.097	0.100	0.609	0.591	3.28	0.783
	303	1.66	0.739	0.101	0.110	0.595	0.547	2.25	1.160
	308	1.64	0.936	0.111	0.120	0.551	0.509	1.75	1.517
	313	1.61	1.133	0.120	0.130	0.517	0.478	1.42	1.899

Table (4-3) : *Corrosion parameters for the polarization of 5083 alloy in aerated NaOH solution pH=13 with presence of NaCl at four temperatures.*

Conc. Of NaCl mol.dm^{-3}	T (K)	Corrosion		b (V.decade^{-1})		α		$R_p/10^3$ (Ω.cm^{-2})	$i_o/10^{-5}$ (A.cm^{-2})
		$-E_{corr}$ (V)	$i_{corr}/10^{-3}$ (A.cm^{-2})	$-b_c$	$+b_a$	$α_c$	$α_a$		
1x10^{-3}	298	2.07	0.184	0.070	0.051	0.845	1.159	11.25	0.228
	303	2.05	0.271	0.091	0.061	0.661	0.985	7.56	0.345
	308	2.03	0.460	0.130	0.080	0.470	0.764	4.41	0.602
	313	2.02	0.644	0.167	0.103	0.372	0.603	3.14	0.858
1x10^{-2}	298	2.06	0.260	0.094	0.067	0.629	0.882	7.92	0.324
	303	2.04	0.456	0.099	0.069	0.607	0.871	4.47	0.584
	308	2.03	0.552	0.101	0.071	0.605	0.861	3.68	0.721
	313	1.99	0.664	0.111	0.080	0.559	0.776	2.99	0.902
0.1	298	2.00	0.606	0.068	0.052	0.869	1.137	3.30	0.778
	303	1.98	0.880	0.081	0.082	0.742	0.732	2.25	1.160
	308	1.95	1.241	0.123	0.111	0.497	0.551	1.57	1.690
	313	1.93	1.349	0.167	0.136	0.372	0.457	1.43	1.886

Table (4-4) : *Corrosion parameters for the polarization of 7075 alloy in aerated NaOH solution pH=13 with presence of NaCl at four temperatures.*

Conc. Of NaCl mol.dm^{-3}	T (K)	Corrosion		b (V.decade^{-1})		α		$R_p/10^3$ (Ω.cm^{-2})	$i_o/10^{-5}$ (A.cm^{-2})
		$-E_{corr}$ (V)	$i_{corr}/10^{-3}$ (A.cm^{-2})	$-b_c$	$+b_a$	$α_c$	$α_a$		
1x10^{-3}	298	2.08	0.010	0.020	0.012	2.956	4.927	208.0	0.012
	303	2.07	0.017	0.030	0.023	2.004	2.614	121.76	0.021
	308	2.04	0.080	0.035	0.030	1.746	2.037	25.5	0.104
	313	2.03	0.093	0.041	0.041	1.515	1.515	21.83	0.124
1x10^{-2}	298	2.07	0.200	0.094	0.052	0.629	1.137	9.95	0.258
	303	2.06	0.260	0.100	0.068	0.601	0.884	7.92	0.329
	308	2.05	0.277	0.112	0.070	0.546	0.873	7.40	0.359
	313	2.02	0.294	0.129	0.080	0.481	0.776	6.87	0.393
0.1	298	2.06	0.346	0.090	0.045	0.657	1.314	5.95	0.431
	303	2.05	0.415	0.093	0.052	0.646	1.156	4.94	0.528
	308	2.04	0.519	0.097	0.065	0.629	0.940	3.93	0.675
	313	2.01	0.588	0.100	0.071	0.621	0.875	3.42	0.789

Table (4-5) : *Corrosion parameters for the polarization of pure aluminium in aerated NaOH solution pH=11 with presence of NaCl at four temperatures.*

Conc. Of NaCl mol.dm^{-3}	T (K)	Corrosion		b (V.decade^{-1})		a		R_p (Ω.cm^{-2})	i_o (A.cm^{-2})
		$-E_{corr}$ (V)	i_{corr} (A.cm^{-2})	$-b_c$	$+b_a$	a_c	a_a		
1x10^{-3}	298	2.12	0.463x10^{-7}	0.018	0.016	3.285	3.695	4.58×10^7	0.56×10^{-9}
	303	2.10	0.588x10^{-7}	0.021	0.017	2.863	3.536	3.57×10^7	0.73×10^{-9}
	308	2.08	0.772x10^{-7}	0.023	0.019	2.657	3.216	2.69×10^7	0.98×10^{-9}
	313	2.03	0.926x10^{-7}	0.028	0.023	2.218	2.699	2.19×10^7	1.23×10^{-9}
1x10^{-2}	298	2.17	0.926x10^{-6}	0.067	0.066	0.882	0.896	2.34×10^6	1.09×10^{-8}
	303	2.16	1.019x10^{-6}	0.079	0.073	0.761	0.823	2.12×10^6	1.23×10^{-8}
	308	2.15	1.543x10^{-6}	0.107	0.086	0.571	0.711	1.39×10^6	1.90×10^{-8}
	313	2.14	1.605x10^{-6}	0.211	0.091	0.294	0.682	1.33×10^6	2.02×10^{-8}
0.1	298	2.13	0.277x10^{-4}	0.055	0.072	1.075	0.821	7.69×10^4	0.33×10^{-6}
	303	2.09	0.432x10^{-4}	0.060	0.088	1.002	0.683	4.84×10^4	0.53×10^{-6}
	308	2.07	0.586x10^{-4}	0.075	0.091	0.815	0.672	3.53×10^4	0.75×10^{-6}
	313	2.05	0.741x10^{-4}	0.094	0.103	0.661	0.603	2.77×10^4	0.97×10^{-6}

Table (4-6) : *Corrosion parameters for the polarization of 2024 alloy in aerated NaOH solution pH=11 with presence of NaCl at four temperatures.*

Conc. Of NaCl mol.dm^{-3}	T (K)	Corrosion		b (V.decade^{-1})		a		R_p/10^5 (Ω.cm^{-2})	i_o/10^{-7} (A.cm^{-2})
		$-E_{corr}$ (V)	i_{corr}/10^{-5} (A.cm^{-2})	$-b_c$	$+b_a$	a_c	a_a		
1x10^{-3}	298	1.43	0.640	0.065	0.063	0.909	0.938	2.23	1.151
	303	1.40	0.739	0.081	0.081	0.742	0.742	0.89	1.381
	308	1.37	0.985	0.091	0.085	0.672	0.719	1.39	1.909
	313	1.35	1.232	0.100	0.098	0.621	0.634	1.10	2.451
1x10^{-2}	298	1.71	0.148	0.053	0.051	1.116	1.159	11.55	0.222
	303	1.54	0.172	0.081	0.063	0.742	0.954	8.95	0.292
	308	1.51	0.197	0.083	0.077	0.736	0.794	7.66	0.346
	313	1.47	0.222	0.092	0.083	0.675	0.748	6.62	0.407
0.1	298	1.50	1.478	0.045	0.058	1.314	1.019	1.01	2.542
	303	1.43	2.217	0.058	0.066	1.036	0.911	0.65	4.015
	308	1.40	3.448	0.071	0.070	0.861	0.873	0.41	6.473
	313	1.34	4.187	0.081	0.075	0.767	0.828	0.32	8.428

Table (4-7) : *Corrosion parameters for the polarization of 5083 alloy in aerated NaOH solution pH=11 with presence of NaCl at four temperatures.*

Conc. Of NaCl mol.dm^{-3}	T (K)	Corrosion		b (V.decade^{-1})		α		$R_p/10^6$ ($\Omega.cm^{-2}$)	$i_o/10^{-8}$ (A.cm^{-2})
		$-E_{corr}$ (V)	$i_{corr}/10^{-6}$ (A.cm^{-2})	$-b_c$	$+b_a$	α_c	α_a		
1x10^{-3}	298	1.83	0.920	0.033	0.025	1.792	2.365	1.99	1.289
	303	1.79	4.580	0.045	0.041	1.336	1.466	0.39	6.692
	308	1.78	5.977	0.055	0.060	1.111	1.018	0.30	8.847
	313	1.76	7.356	0.059	0.075	1.053	0.828	0.24	11.237
1x10^{-2}	298	1.95	1.011	0.048	0.075	1.232	0.788	1.93	1.330
	303	1.94	1.511	0.053	0.100	1.134	0.601	1.28	2.039
	308	1.93	1.931	0.091	0.104	0.672	0.588	1.01	2.627
	313	1.91	2.069	0.115	0.107	0.539	0.580	0.92	2.931
0.1	298	1.90	0.382	0.040	0.050	1.478	1.182	4.97	0.516
	303	1.87	0.414	0.050	0.067	1.202	0.897	4.52	0.577
	308	1.86	0.460	0.053	0.091	1.153	0.672	4.04	0.657
	313	1.85	0.552	0.062	0.103	1.002	0.603	3.35	0.805

Table (4-8) : *Corrosion parameters for the polarization of 7075 alloy in aerated NaOH solution pH=11 with presence of NaCl at four temperatures.*

Conc. Of NaCl mol.dm^{-3}	T (K)	Corrosion		b (V.decade^{-1})		α		$R_p/10^6$ ($\Omega.cm^{-2}$)	$i_o/10^{-8}$ (A.cm^{-2})
		$-E_{corr}$ (V)	$i_{corr}/10^{-6}$ (A.cm^{-2})	$-b_c$	$+b_a$	α_c	α_a		
1x10^{-3}	298	1.89	0.081	0.028	0.020	2.112	2.956	23.33	0.110
	303	1.88	0.087	0.035	0.023	1.718	2.614	21.61	0.121
	308	1.84	0.121	0.040	0.029	1.528	2.107	15.21	0.174
	313	1.82	0.225	0.048	0.035	1.294	1.774	8.09	0.333
1x10^{-2}	298	2.03	0.761	0.046	0.054	1.285	1.095	2.67	0.961
	303	1.98	0.865	0.054	0.056	1.113	1.073	2.29	1.139
	308	1.95	1.038	0.058	0.065	1.054	0.940	1.88	1.412
	313	1.92	1.142	0.071	0.073	0.875	0.851	1.68	1.605
0.1	298	1.91	0.239	0.040	0.136	1.478	0.435	7.99	0.321
	303	1.89	0.273	0.057	0.150	1.055	0.401	6.92	0.377
	308	1.88	0.433	0.075	0.188	0.815	0.325	4.34	0.612
	313	1.87	0.657	0.120	0.250	0.517	0.248	2.85	0.946

Table (4-9) : Corrosion parameters for the polarization of pure aluminium in aerated NaOH solution pH=9 with presence of NaCl at four temperatures.

Conc. NaCl mol.dm^{-3}	T (K)	Corrosion		b (V.decade^{-1})		α		$R_p/10^7$ Ω.cm^{-2}	$i_o/10^{-9}$ A.cm^{-2}	Passivity	
		$-E_{corr}$ (V)	$i_{corr}/10^{-7}$ A.cm^{-2}	$-b_c$	$+b_a$	$α_c$	$α_a$			$-E_p$ (V)	$i_p/10^{-5}$ A.cm^{-2}
1x10^{-3}	298	2.24	12.346	0.048	-	1.232	-	0.18	14.26	0.87	0.444
	303	2.19	15.100	0.100	-	0.601	-	0.15	17.40	0.85	0.463
	308	2.14	18.519	0.214	-	0.286	-	0.12	22.11	0.81	0.501
	313	1.93	23.148	0.300	-	0.207	-	0.08	33.71	0.69	0.525
1x10^{-2}	298	0.79	0.420	0.301	0.300	0.196	0.197	1.88	1.365	-	-
	303	0.73	0.432	0.333	0.380	0.181	0.158	1.69	1.544	-	-
	308	0.71	0.448	0.354	0.470	0.173	0.130	1.58	1.679	-	-
	313	0.69	0.457	0.375	0.510	0.166	0.122	1.51	1.786	-	-
0.1	298	0.71	0.401	0.185	0.190	0.319	0.311	1.77	1.450	-	-
	303	0.72	0.586	0.174	0.188	0.345	0.319	1.23	2.122	-	-
	308	0.73	0.648	0.167	0.136	0.366	0.449	1.13	2.349	-	-
	313	0.77	0.772	0.115	0.120	0.539	0.517	1.00	2.697	-	-

Table (4-10) : Corrosion parameters for the polarization of 2024 alloy in aerated NaOH solution pH=9 with presence of NaCl at four temperatures.

Conc. NaCl mol.dm^{-3}	T (K)	Corrosion		b (V.decade^{-1})		α		$R_p/10^7$ Ω.cm^{-2}	$i_o/10^{-9}$ A.cm^{-2}	Passivity	
		$-E_{corr}$ (V)	$i_{corr}/10^{-7}$ A.cm^{-2}	$-b_c$	$+b_a$	$α_c$	$α_a$			$-E_p$ (V)	$i_p/10^{-6}$ A.cm^{-2}
1x10^{-3}	298	1.40	9.852	0.052	-	1.137	-	0.14	18.33	0.43	1.970
	303	1.35	10.837	0.078	-	0.771	-	0.12	21.75	0.41	2.069
	308	1.26	11.823	0.081	-	0.754	-	0.11	24.13	0.39	2.118
	313	1.22	12.510	0.099	-	0.627	-	0.10	26.97	0.38	2.217
1x10^{-2}	298	-0.01	0.788	0.375	0.251	0.158	0.236	0.01	256.7	-	-
	303	-0.03	0.837	0.176	0.262	0.342	0.229	0.04	65.25	-	-
	308	-0.06	0.887	0.143	0.276	0.427	0.221	0.07	37.91	-	-
	313	-0.07	1.182	0.097	0.301	0.640	0.206	0.06	44.95	-	-
0.1	298	0.14	0.788	0.056	0.201	1.056	0.294	0.18	14.26	-	-
	303	0.15	0.985	0.056	0.192	1.073	0.313	0.15	17.40	-	-
	308	0.16	1.626	0.038	0.185	1.608	0.330	0.10	26.54	-	-
	313	0.17	2.956	0.034	0.177	1.826	0.351	0.06	44.95	-	-

Table (4-11) : Corrosion parameters for the polarization of 5083 alloy in aerated NaOH solution pH=9 with presence of NaCl at four temperatures.

Conc. NaCl mol.dm^{-3}	T (K)	Corrosion		b (V.decade^{-1})		α		$R_p/10^7$ Ω.cm^{-2}	$i_d/10^{-9}$ A.cm^{-2}	Passivity	
		$-E_{corr}$ (V)	i_{corr} /10^{-7} A.cm^{-2}	$-b_c$	$+b_a$	$α_c$	$α_a$			$-E_p$ (V)	$i_p/10^{-5}$ A.cm^{-2}
1x10^{-3}	298	1.80	21.820	0.042	-	1.408	-	0.08	32.08	0.50	0.552
	303	1.71	22.988	0.065	-	0.925	-	0.07	37.28	0.44	0.598
	308	1.65	23.908	0.081	-	0.754	-	0.06	44.23	0.42	0.667
	313	1.50	27.126	0.101	-	0.615	-	0.05	53.94	0.40	0.736
1x10^{-2}	298	0.71	0.690	0.330	0.300	0.179	0.197	1.03	2.492	-	-
	303	0.69	0.713	0.300	0.250	0.200	0.240	0.97	2.691	-	-
	308	0.64	0.736	0.273	0.234	0.224	0.261	0.87	3.050	-	-
	313	0.63	0.759	0.214	0.214	0.290	0.290	0.83	3.249	-	-
0.1	298	0.79	0.828	0.231	0.230	0.256	0.257	0.95	2.702	-	-
	303	0.80	0.966	0.220	0.215	0.273	0.279	0.83	3.145	-	-
	308	0.83	1.103	0.214	0.188	0.286	0.325	0.75	3.538	-	-
	313	0.88	1.517	0.188	0.136	0.330	0.457	0.58	4.650	-	-

Table (4-12) : Corrosion parameters for the polarization of 7075 alloy in aerated NaOH solution pH=9 with presence of NaCl at four temperatures.

Conc. NaCl mol.dm^{-3}	T (K)	Corrosion		b (V.decade^{-1})		α		$R_p/10^6$ Ω.cm^{-2}	$i_d/10^{-8}$ A.cm^{-2}	Passivity	
		$-E_{corr}$ (V)	i_{corr} /10^{-6} A.cm^{-2}	$-b_c$	$+b_a$	$α_c$	$α_a$			$-E_p$ (V)	$i_p/10^{-5}$ A.cm^{-2}
1x10^{-3}	298	2.02	1.107	0.188	-	0.314	-	1.82	1.410	1.25	0.588
	303	2.00	1.799	0.214	-	0.281	-	1.11	2.351	1.21	0.602
	308	1.92	2.422	0.273	-	0.224	-	0.79	3.359	1.18	0.612
	313	1.85	2.768	0.550	-	0.113	-	0.67	4.025	1.16	0.623
1x10^{-2}	298	2.15	1.384	0.022	-	2.687	-	1.55	1.656	1.45	0.450
	303	2.13	2.076	0.063	-	0.954	-	1.03	2.534	1.42	0.554
	308	1.84	3.806	0.107	-	0.571	-	0.48	5.529	1.24	0.657
	313	1.76	4.498	0.150	-	0.414	-	0.39	6.915	1.18	0.757
0.1	298	1.06	0.083	0.045	0.059	1.314	1.002	12.77	0.201	-	-
	303	1.07	0.087	0.111	0.103	0.542	0.584	12.30	0.212	-	-
	308	1.19	0.100	0.167	0.120	0.366	0.509	11.90	0.223	-	-
	313	1.21	0.138	0.178	0.130	0.349	0.478	8.77	0.307	-	-

4-4 Corrosion Potential and Corrosion Current Density

The rate of corrosion increased in the presence of chloride ion (at three experimental concentrations of NaCl) and with the increasing temperature due to the exposure of the metal (and alloys) surface to the aggressive solution. The results can be tentatively explained on the basis of the occurrence of competitive adsorption of chloride ion on the metal or alloy surface. The role of chloride ions depends on the concentration of medium and NaCl additive. Generally, at pH=13 the presence of chloride ions caused a shift in the corrosion potentials (E_{corr}) to more negative values and the corrosion current densities (i_{corr}) to higher values in the comparison with the absence of chloride ions in solution.

At pH=11, the presence of chloride ions in solution shifts the (E_{corr}) to more negative values and (i_{corr}) to lower value for pure Al and its alloys except the Al-Cu-Mg alloy. The presence of chloride ions in solution at pH=9 shifts the (E_{corr}) either in active or noble direction according to concentration of NaCl which presence in solution and shifts the (i_{corr}) to lower values for pure Al and its alloys except the Al-Zn-Mg alloy.

The data which have been given in Tables (4-1) to (4-12), can be summarized in the following :

_A- In pH = 13_

_1-_ At constant concentration of NaCl and temperature, (E_{corr}) changes according to the following sequence as shown in Fig. (4-8 a, b, c) at three concentrations of NaCl :   *(E_corr) values*   pure *Al   7075   5083   2024*

Noble Direction

_2-_ At constant concentration of NaCl and temperature, Fig. (4-9 a, b, c) show the effect of chloride ion on the (i_{corr}) values for pure Al and its alloys, where the (i_{corr}) values take the following sequences:

(i_corr) values   pure *Al > 5083 alloy > 2024 alloy > 7075 alloy*

117

3- To the same electrode at a constant temperature, the increase of concentration of NaCl shifts the (E_{corr}) toward the noble direction as shown in Fig. (4-10) for pure Al and its alloys.

4- To the same electrode at a constant temperature, (i_{corr}) increases with increasing the concentration of NaCl as shown in Fig. (4-11) for pure Al and its alloys.

B- In pH = 11

1- At constant concentration of NaCl and temperature, (E_{corr}) changes according to the following sequences and shown in Fig. (4-12 a, b, c) at three concentrations of NaCl :

$$\xrightarrow{\hspace{3cm}}$$
Noble Direction

(E_{corr}) *values*   *pure Al,   7075,   5083,   2024*

2- At a constant temperature, (i_{corr}) varies with the variation of the concentration for pure Al and its alloy and behaves different behaviour as follow:

	at $10^{-3}M$ Cl^-	*2024 > 5083 > 7075 > pure Al*
i_{corr}	*at* $10^{-2}M$ Cl^-	*2024 > 5083 > pure Al > 7075*
	at $0.1\ M\ Cl^-$	*pure Al > 2024 > 5083 > 7075*

This sequence indicates the sensitivity of pure Al to chloride ions in 1×10^{-3} mol.dm^{-3} NaOH solution (pH =11).

3- To the same electrode at a constant temperature, Fig. (4-13) show the variation of (E_{corr}) with the concentration of NaCl for pure Al and its alloys.

4- To the same electrode at a constant temperature, (i_{corr}) gives a different behaviour with the increasing of the concentration of NaCl as shown in Fig. (4-14) for pure Al and its alloys.

C- In pH = 9

(E_{corr}) and (i_{corr}) give a different behaviour for pure Al and its alloys in this value of pH (because of very small concentration of medium), also there are no certain sequence in this case neither at a constant concentration of NaCl nor at a constant temperature.

_D-_ Generally, the increase of temperature shift the (E_{corr}) toward the noble direction except several cases, and leads to increase of (i_{corr}) in all cases.

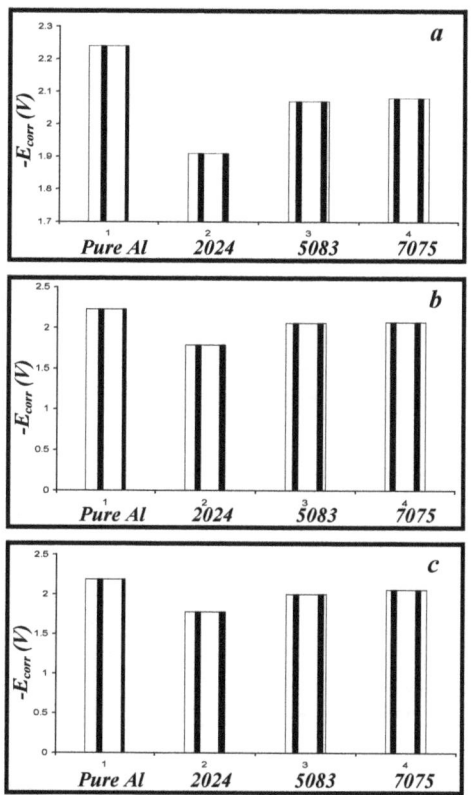

Fig. (4-8) _: Values of (E_{corr}) plotted for pure Al and its alloys in pH=13 at 298K in the presence of NaCl_
a: 10^{-3} mol.dm^{-3}, b: 10^{-2} mol.dm^{-3}, c: 0.1 mol.dm^{-3}.

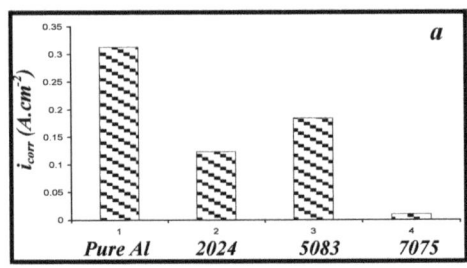

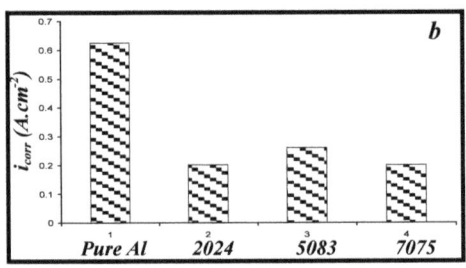

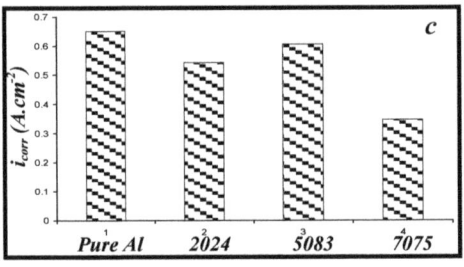

Fig. (4-9) : *Values of (i_{corr}) plotted for pure Al and its*
alloys in pH=13 at 298K in the presence of NaCl
a: 10^{-3} mol.dm^{-3}, b: 10^{-2} mol.dm^{-3}, c: 0.1 mol.dm^{-3}.

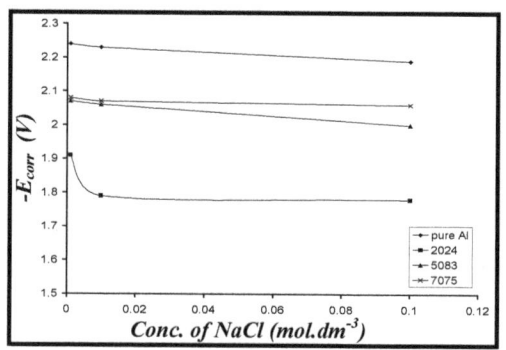

Fig. (4-10) : *Effect of concentration of NaCl on*
the values of (E_{corr}) for pure Al and
its alloys in pH=13.

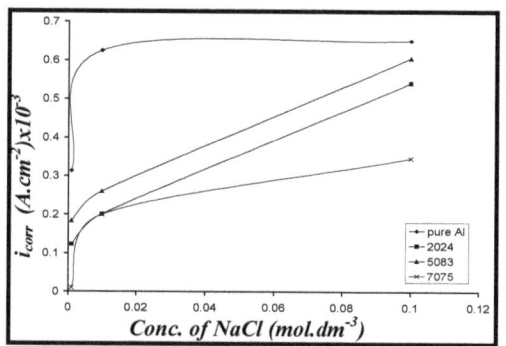

Fig. (4-11) : *Effect of concentration of NaCl on*
the values of (i_{corr}) for pure Al and
its alloys in pH=13.

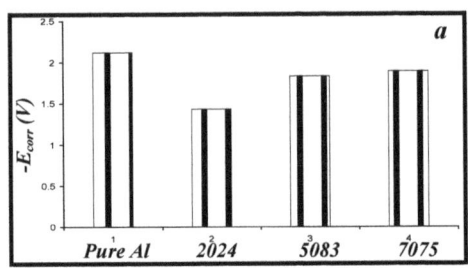

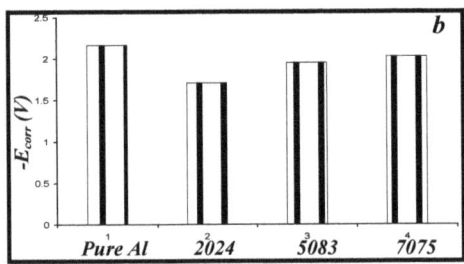

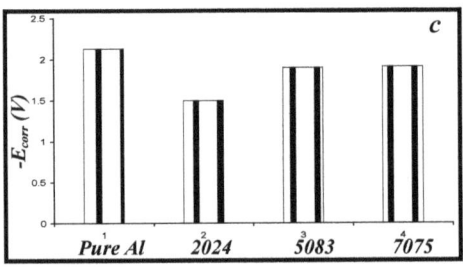

Fig. (4-12) *: Values of (E_{corr}) plotted for pure*

Al and its alloys in pH=11 at 298K in the

presence of NaCl

a:1x10⁻³ mol.dm⁻³, b:1x10⁻² mol.dm⁻³, c: 0.1 mol.dm⁻³.

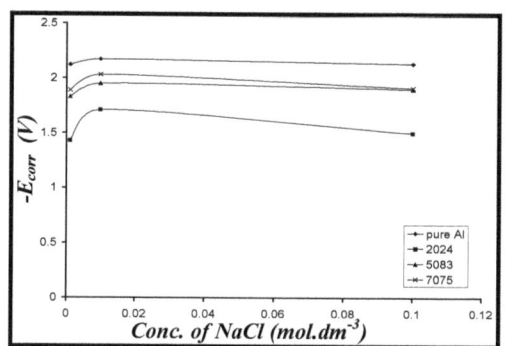

***Fig. (4-13)* :** *Effect of concentration of NaCl on*
the values of (E$_{corr}$) for pure Al and
its alloys in pH=11.

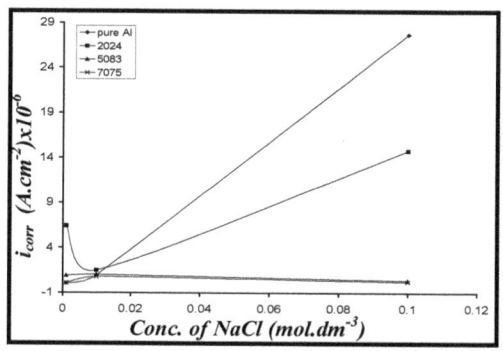

***Fig. (4-14)* :** *Effect of concentration of NaCl on*
the values of (i$_{corr}$) for pure Al and
its alloys in pH=11.

4-5 Tafel Slope (b) and Transfer Coefficient (α)

Tables (4-1) to (4-12) show the influence of temperature and concentration of NaCl (additive) on the cathodic (b_c) and anodic (b_a) Tafel slopes which have been obtained from the polarization curves of pure Al and its alloys in NaOH solution (pH 13, 11, and 9) over the temperature range (298 – 313)K.

Values of the transfer coefficients for cathodic ($α_c$) and anodic ($α_a$) processes have been calculated from the corresponding values of the Tafel slope (b) using the relations in (3-1) and (3-2) equations[125, 35].

Generally, in pH = 13 and 11 values of b < 0.120 for pure Al and its alloys, while in pH = 9 most cases b > 0.120. The results of the tables indicate that the variation of the Tafel slopes and of the corresponding transfer coefficients could be interpreted in terms of the variation of the rate – determining step from charge transfer process to either chemical – desorption or to electrochemical desorption.

The variation of the anodic transfer coefficients ($α_a$) may be attributed to the variation of the rate – determining step in the metal dissolution reaction. A change in the mechanism as well as in the rate – determining step, cannot be ignored throughout the anodic processes.

4-6 Polarization Resistance

The term ($\frac{E_{corr}}{i_{corr}}$) corresponds to the resistance (R) of the metal/solution interface to charge – transfer reaction. It is also a measure of the resistance of the metal to corrosion in the solution in which the metal is immersed. The reaction resistance (R_p), which mainly depends upon the equilibrium exchange current density (i_o) determines what may be termed the polarizability, i.e., what overpotential ($\eta=E-E_{corr}$) a particular current density needs (for a driven cell) or produces (for a spontaneously performing cell).

Tables (4-1) to (4-12) indicates the results of polarization resistance of pure Al and its alloys as follow:

In pH=13   *7075 > 2024 > 5083 > pure Al*

In pH=11 (1x10^{-3} and 1x10^{-2} mol.dm^{-3} Cl$^-$)   *pure Al > 7075 > 5083 > 2024*

         (0.1 mol.dm^{-3} Cl$^-$)   *7075 > 5083 > 2024 > pure Al*

In pH=9 (1x10^{-3} mol.dm^{-3} Cl$^-$)   *pure Al > 2024 > 5083 > 7075*

        (1x10^{-2} and 0.1 mol.dm^{-3} Cl$^-$)   *pure Al > 5083 > 2024 > 7075*

While in the absence of chloride ions in NaOH solution at three values of pH (13, 11, and 9) the results of polarization resistance are:

Pure Al >>   7075 > 5083 > 2024

Chapter Five: Results and Discussion of Corrosion Inhibition

5-1 Introduction (Corrosion Control)

There are two effective methods to protect aluminium alloys from corrosion :

5-1-1 Cathodic Protection

1- Use of an external source to reduce the electrode potential.

2- Galvanic coupling : Aluminium can be protected cathodically by galvanic coupling to an active alloy such as plating Al – Zn alloy onto pure aluminium, pits will penetrate the cladding only while the base metal is protected[155].

5-1-2 Inhibitors

Lorking and Mayne (1961)[155] classified the anions into three categories according to their action on aluminium.

I- Anions not forming complexes:

a- Non – oxidizing anions such as benzoate, phosphate, and acetate which inhibit corrosion within the neutral range of pH.

b- Oxidizing anions such as chromate and nitrate which inhibit corrosion in wider range of pH than class (a).

II- Anions forming soluble complexes with aluminium such as citrate and tartrate. The corrosion rate in the presence of such anions is higher than solution of class (I)

III- Anions forming soluble complexes such as chlorides, causing corrosion of aluminium in neutral solution.

Corrosion of aluminium has been a subject of numerous studies due to the importance of this material in contemporary civilization. It is well known that there is a potential region in which the rate of corrosion is relatively small, of the order $1\mu A.cm^{-2}$ (or 3×10^{-4} $mg.cm^{-2}$ h^{-1} in material equivalent). In the absence of air or other 'depolarizer', or of an external electric field, aluminium in solution rests within this 'potential window' and, hence, is a very stable material. Yet, this window is limited on both the anodic and the cathodic sides. At potentials of about (-0.8V) with respect to the saturated calomel electrode (SCE) a sharp rise in anodic current occurs, with a resulting localized dissolution (pitting corrosion)[89].

On the cathodic side a similar phenomenon is observed. If the potential is driven negatively beyond about (-1.4 V/SCE) a sharp rise in current occurs, not only producing gaseous hydrogen but also causing dissolution of the metal by the chemical attack of hydroxyl ions formed in the same process[89].

The interesting fact is that this dissolution up to three times more efficient than that on the anodic side when calculated in terms of current efficiency. One aluminium atom is dissolved for each electron exchanged in the cathodic process, because the process consists of the reactions :

$$H_2O + e \leftrightarrow H + OH^- \quad ...(1)$$

and

$$Al + OH^- + H_2O \leftrightarrow AlO_2^- + 3/2\ H_2 \quad ...(2)$$

Thus, the production and participation of OH^- ions is essential for the "cathodic corrosion of aluminium". Cathodic corrosion can be equally damaging as anodic in situations where the potential is driven negative[89].

Hence, the possible inhibitive action of the different substance is sought. This work an attempt to inhibit corrosion of pure Al and its three alloys by using sodium acetate (as organic inhibitor) and sodium chromate (as inorganic inhibitor) in the basic media (with three different pH 13, 11, and 9) at four temperature in the range (298 – 313)K.

5-2 Results Of Polarization Curves

5-2-1 Addition Of Sodium Acetate

The addition of sodium acetate (CH_3COONa) as an organic inhibitor to 0.1 mol.dm^{-3} NaOH solution (pH=13) does not change polarization curves shape for pure Al which is similar to that obtained for Al alloys (2024, 5083, and 7075) in presence of (0.05 mol.dm^{-3}) sodium acetate as shown in Fig. (5-1). The same polarization curve is observed in the presence of a higher concentration of sodium acetate (0.10 and 0.15 mol.dm^{-3}) in pH=13 at four temperatures for pure aluminium and its alloys.

The polarization curve mainly consists of the cathodic and anodic Tafel region. At three concentrations of sodium acetate, the inhibitor shifts the (E_{corr}) in the more negative (active) direction and shift (i_{corr}) to a larger value when compared with its values without the addition of this inhibitor.

The similar influence of the inhibitor (CH_3COONa) is observed in the study of polarization curve of pure Al and its alloys with three concentrations of this inhibitor at four temperatures in the 1×10^{-3} mol.dm^{-3} NaOH solution (pH=11). Fig. (5-2) show this influence for 7075 alloy in pH=11 with the presence of 0.1 mol.dm^{-3} CH_3COONa at 298K.

The addition of sodium acetate to the 1×10^{-5} mol.dm^{-3} NaOH solution (pH=9) produces major changes in the Tafel plot (polarization curve). Figure (5-3) shows this change at a low concentration of sodium acetate addition (0.05 mol.dm^{-3} CH_3COONa) for 2024 alloy in pH=9 at 298 K which is similar to that obtained for pure Al and other alloys, in general the polarization curve consists of two sections, the wide cathodic Tafel (abc section) and the wide anodic Tafel (cde section) region.

The presence of sodium acetate with this concentration breaks down any passive layer may be produced. The presence higher concentration of

129

CH$_3$COONa (0.1 and 0.15 mol.dm^{-3}) in pH=9 gives another shape of the polarization curve as shown in Fig. (5-4).

The section (abc) represents the cathodic Tafel region. Anodic dissolution of metal begins at point (c) and continues along (cd). Along the section (def) the metal hydroxide is expected to be formed. The hydroxide soon dissociates into metal oxide (Al$_2$O$_3$) on aluminium surface which behaves as passive layer (protective film). The breakdown of passivity begins at point (f) and continues along (fg).

Tables (5-1) to (5-12) show the results of the polarization curves for pure Al and its alloys in NaOH solution (pH=13, 11, and 9) in the presence of sodium acetate with three concentrations at four temperatures.

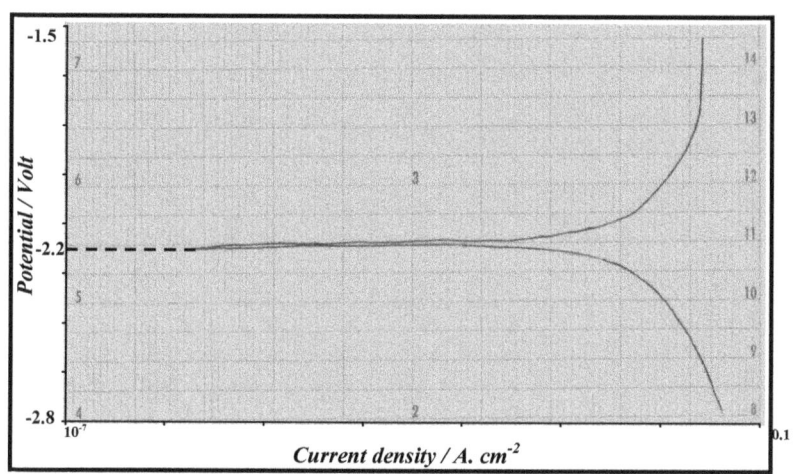

Fig. (5-1) : *The polarization curve of pure Al in pH=13*

in presence of 0.05 mol.dm⁻³ CH₃COONa at 298K.

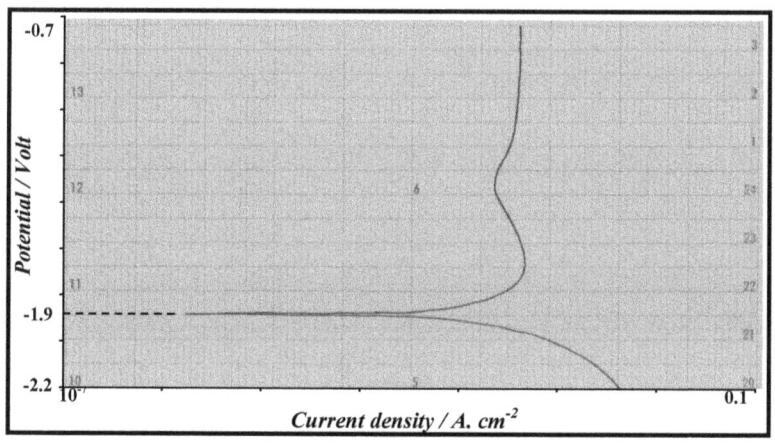

Fig. (5-2) : *The polarization curve of 7075 alloy in pH=11*

in presence of 0.1 mol.dm⁻³ CH₃COONa at 298K.

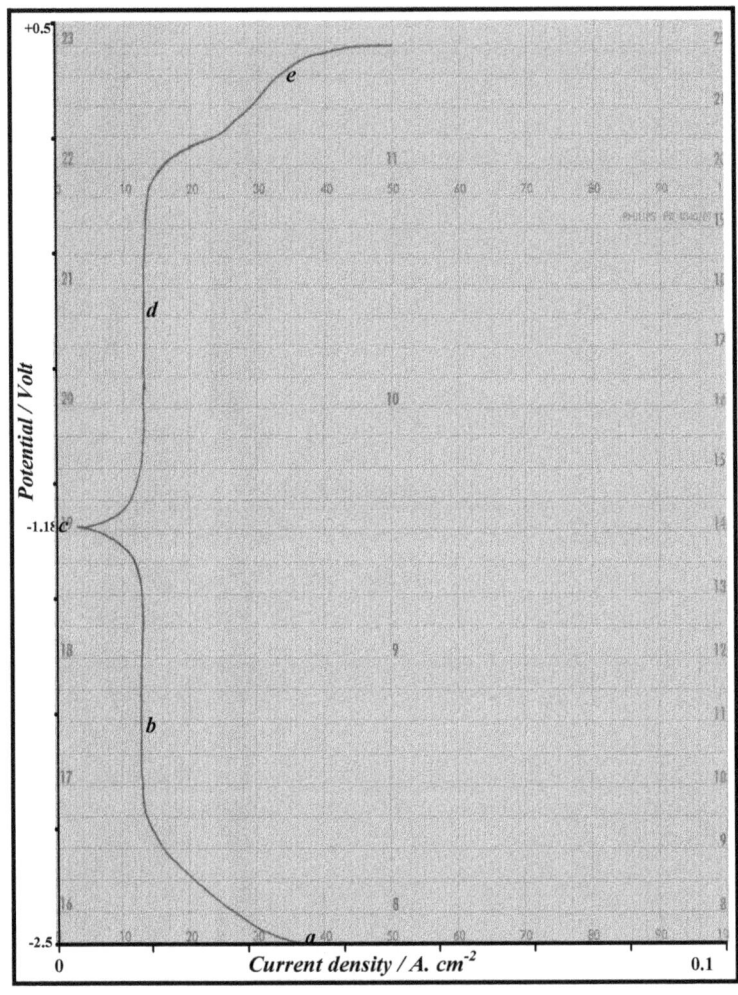

Fig. (5-3) _: The polarization curve of 2024 alloy in pH=9_
in presence of (0.05 mol.dm^{-3}) CH$_3$COONa at 298K.

132

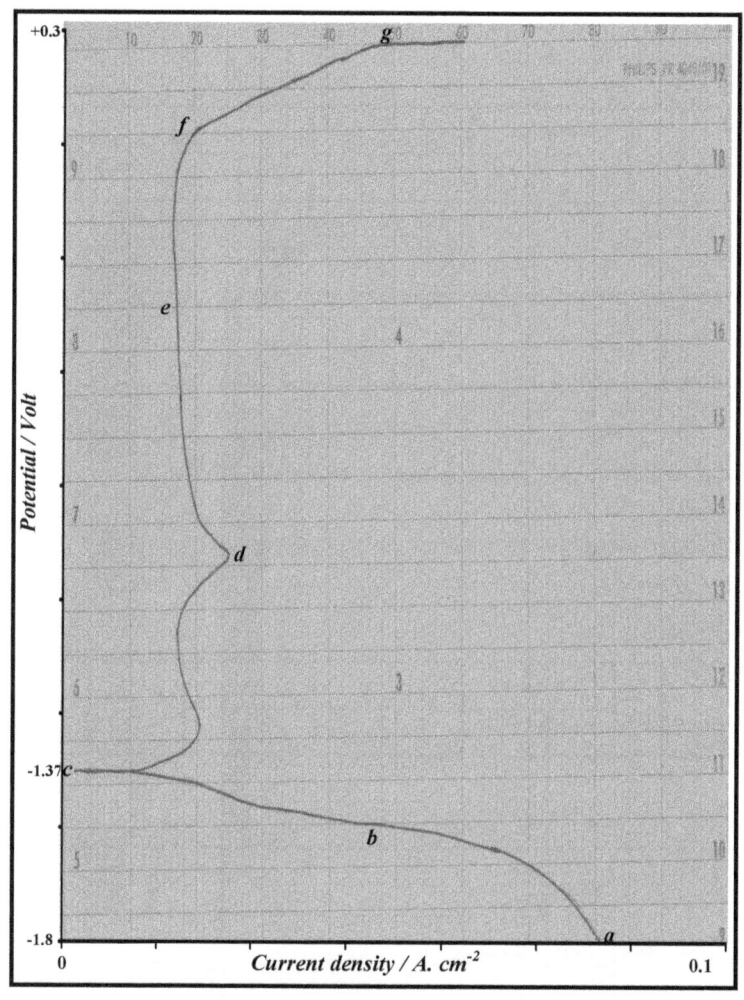

Fig. (5-4) *: The polarization curve of pure Al in pH=9*

in presence of 0.10 mol.dm⁻³

CH₃COONa at 298K.

133

Table (5-1) : *Corrosion parameters for the polarization of pure aluminium in aerated NaOH solution pH=13 with presence of CH_3COONa at four temperatures.*

Conc. Of CH_3COONa mol.dm^{-3}	T (K)	Corrosion		b (V.decade^{-1})		α		$R_r/10^3$ (Ω.cm^{-2})	$i_d/10^{-5}$ (A.cm^{-2})
		$-E_{corr}$ (V)	$i_{corr}/10^{-3}$ (A.cm^{-2})	$-b_c$	$+b_a$	α_c	α_a		
0.05	298	2.20	0.432	0.083	0.079	0.71	0.75	5.09	0.504
	303	2.19	0.486	0.095	0.085	0.63	0.71	4.51	0.579
	308	2.18	0.579	0.107	0.091	0.57	0.67	3.77	0.704
	313	2.17	0.895	0.125	0.100	0.50	0.62	2.42	1.115
0.10	298	2.19	0.246	0.085	0.055	0.70	1.07	8.86	0.290
	303	2.18	0.308	0.090	0.076	0.67	0.79	7.08	0.369
	308	2.17	0.468	0.100	0.081	0.61	0.75	4.64	0.572
	313	2.16	0.639	0.120	0.092	0.52	0.67	3.38	0.798
0.15	298	2.24	0.442	0.094	0.066	0.63	0.90	5.07	0.506
	303	2.23	0.535	0.103	0.078	0.58	0.77	4.17	0.626
	308	2.21	0.555	0.120	0.085	0.51	0.72	3.98	0.667
	313	2.18	0.617	0.127	0.091	0.49	0.68	3.53	0.764

Table (5-2) : *Corrosion parameters for the polarization of 2024 alloy in aerated NaOH solution pH=13 with presence of CH_3COONa at four temperatures.*

Conc. Of CH_3COONa mol.dm^{-3}	T (K)	Corrosion		b (V.decade^{-1})		α		$R_r/10^3$ (Ω.cm^{-2})	$i_d/10^{-5}$ (A.cm^{-2})
		$-E_{corr}$ (V)	$i_{corr}/10^{-3}$ (A.cm^{-2})	$-b_c$	$+b_a$	α_c	α_a		
0.05	298	1.82	0.487	0.083	0.076	0.71	0.77	3.74	0.686
	303	1.76	0.519	0.089	0.090	0.67	0.66	3.39	0.770
	308	1.69	0.588	0.096	0.096	0.63	0.63	2.87	0.925
	313	1.65	0.837	0.115	0.111	0.54	0.56	1.97	1.369
0.10	298	1.81	0.344	0.078	0.075	0.75	0.79	5.26	0.488
	303	1.79	0.443	0.090	0.081	0.66	0.74	4.04	0.646
	308	1.76	0.541	0.096	0.103	0.63	0.59	3.25	0.817
	313	1.71	0.640	0.099	0.115	0.62	0.54	2.67	1.010
0.15	298	1.80	0.454	0.088	0.075	0.67	0.79	3.96	0.648
	303	1.76	0.542	0.090	0.090	0.66	0.67	3.25	0.803
	308	1.71	0.638	0.100	0.115	0.61	0.53	2.68	0.990
	313	1.65	0.788	0.136	0.272	0.46	0.23	2.09	1.290

Table (5-3) _:Corrosion parameters for the polarization of 5083 alloy in aerated NaOH solution pH=13 with presence of CH_3COONa at four temperatures._

Conc. Of CH_3COONa $mol.dm^{-3}$	T (K)	Corrosion		b ($V.decade^{-1}$)		α		$R_p/10^3$ ($\Omega.cm^{-2}$)	$i_o/10^{-5}$ ($A.cm^{-2}$)
		$-E_{corr}$ (V)	$i_{corr}/10^{-3}$ ($A.cm^{-2}$)	$-b_c$	$+b_a$	α_c	α_a		
0.05	298	2.06	0.505	0.075	0.071	0.79	0.83	4.08	0.629
	303	2.05	0.551	0.085	0.075	0.71	0.80	3.72	0.702
	308	2.04	0.597	0.111	0.081	0.55	0.75	3.42	0.776
	313	2.03	0.689	0.150	0.150	0.41	0.41	2.95	0.914
0.10	298	2.06	0.597	0.081	0.055	0.73	1.07	3.45	0.744
	303	2.05	0.735	0.120	0.062	0.50	0.96	2.79	0.935
	308	2.04	0.873	0.136	0.065	0.45	0.94	2.34	1.134
	313	2.03	0.919	0.166	0.136	0.37	0.46	2.21	1.220
0.15	298	2.03	0.474	0.0545	0.050	1.08	1.18	4.28	0.599
	303	2.04	0.571	0.071	0.057	0.84	1.04	3.57	0.731
	308	2.06	0.689	0.081	0.066	0.75	0.92	2.99	0.888
	313	2.07	1.057	0.125	0.083	0.50	0.75	1.96	1.376

Table (5-4) _: Corrosion parameters for the polarization of 7075 alloy in aerated NaOH solution pH=13 with presence of CH_3COONa at four temperatures._

Conc. Of CH_3COONa $mol.dm^{-3}$	T (K)	Corrosion		b ($V.decade^{-1}$)		α		$R_p/10^3$ ($\Omega.cm^{-2}$)	$i_o/10^{-5}$ ($A.cm^{-2}$)
		$-E_{corr}$ (V)	$i_{corr}/10^{-3}$ ($A.cm^{-2}$)	$-b_c$	$+b_a$	α_c	α_a		
0.05	298	2.09	0.311	0.066	0.063	0.89	0.93	6.72	0.382
	303	2.08	0.346	0.100	0.090	0.60	0.66	6.01	0.434
	308	2.06	0.415	0.103	0.120	0.59	0.51	4.96	0.535
	313	2.04	0.449	0.121	0.150	0.51	0.41	4.54	0.594
0.10	298	2.12	0.148	0.062	0.051	0.95	1.14	14.32	0.179
	303	2.11	0.173	0.081	0.066	0.74	0.90	12.20	0.214
	308	2.08	0.449	0.136	0.078	0.45	0.77	4.63	0.573
	313	2.07	0.622	0.142	0.115	0.43	0.54	3.33	0.810
0.15	298	2.13	0.439	0.043	0.046	1.36	1.26	4.85	0.529
	303	2.11	0.523	0.052	0.048	1.14	1.24	4.03	0.648
	308	2.09	0.533	0.055	0.057	1.10	1.06	3.92	0.677
	313	2.08	0.571	0.071	0.068	0.87	0.91	3.64	0.741

Table (5-5) : *Corrosion parameters for the polarization of pure aluminium in aerated NaOH solution pH=11 with presence of CH_3COONa at four temperatures.*

Conc. Of CH_3COO Na mol.dm^{-3}	T (K)	Corrosion		b (V.decade^{-1})		α		$R_p/10^5$ ($\Omega.cm^{-2}$)	$i_d/10^{-7}$ ($A.cm^{-2}$)
		$-E_{corr}$ (V)	$i_{corr}/10^{-5}$ ($A.cm^{-2}$)	$-b_c$	$+b_a$	α_c	α_a		
0.05	298	2.07	0.833	0.046	0.062	1.28	0.95	2.48	1.035
	303	2.06	1.234	0.059	0.082	1.02	0.73	1.67	1.563
	308	2.05	1.443	0.071	0.100	0.86	0.61	1.42	1.869
	313	2.02	1.697	0.085	0.115	0.72	0.54	1.19	2.266
0.10	298	2.18	1.251	0.075	0.076	0.79	0.77	1.74	1.475
	303	2.16	1.698	0.078	0.083	0.76	0.72	1.27	2.055
	308	2.12	2.314	0.085	0.099	0.71	0.62	0.92	2.885
	313	2.05	2.623	0.096	0.130	0.64	0.48	0.78	3.458
0.15	298	2.15	1.434	0.050	0.056	1.16	1.04	1.50	1.711
	303	2.14	1.951	0.060	0.073	1.00	0.82	1.10	2.373
	308	2.13	2.469	0.066	0.100	0.92	0.61	0.86	3.086
	313	2.12	3.703	0.075	0.120	0.83	0.52	0.57	4.732

Table (5-6) : *Corrosion parameters for the polarization of 2024 alloy in aerated NaOH solution pH=11 with presence of CH_3COONa at four temperatures.*

Conc. Of CH_3COO Na mol.dm^{-3}	T (K)	Corrosion		b (V.decade^{-1})		α		$R_p/10^5$ ($\Omega.cm^{-2}$)	$i_d/10^{-7}$ ($A.cm^{-2}$)
		$-E_{corr}$ (V)	$i_{corr}/10^{-5}$ ($A.cm^{-2}$)	$-b_c$	$+b_a$	α_c	α_a		
0.05	298	1.59	0.459	0.037	0.044	1.58	1.34	3.46	0.742
	303	1.56	0.689	0.056	0.048	1.06	1.24	2.26	1.155
	308	1.50	0.837	0.057	0.065	1.05	0.94	1.79	1.483
	313	1.47	1.222	0.060	0.068	1.03	0.91	1.20	2.248
0.10	298	1.51	1.133	0.061	0.075	0.97	0.79	1.33	1.930
	303	1.50	1.370	0.079	0.088	0.76	0.68	1.09	2.394
	308	1.48	1.971	0.096	0.096	0.63	0.63	0.75	3.539
	313	1.43	2.374	0.100	0.100	0.62	0.62	0.60	4.495
0.15	298	1.46	1.403	0.083	0.066	0.71	0.89	1.04	2.468
	303	1.45	1.854	0.093	0.075	0.64	0.80	0.78	3.346
	308	1.41	2.211	0.111	0.078	0.55	0.77	0.64	4.147
	313	1.37	3.374	0.122	0.085	0.51	0.72	0.41	6.578

Table (5-7) : *Corrosion parameters for the polarization of 5083 alloy in aerated NaOH solution pH=11 with presence of CH_3COONa at four temperatures.*

Conc. Of CH_3COONa mol.dm^{-3}	T (K)	Corrosion		b (V.decade^{-1})		α		$R_p/10^5$ ($\Omega.cm^{-2}$)	$i_c/10^{-7}$ ($A.cm^{-2}$)
		$-E_{corr}$ (V)	$i_{corr}/10^{-5}$ ($A.cm^{-2}$)	$-b_c$	$+b_a$	α_c	α_a		
0.05	298	1.86	2.068	0.050	0.075	1.18	0.79	0.90	2.852
	303	1.84	2.206	0.055	0.085	1.08	0.70	0.83	3.145
	308	1.82	2.298	0.061	0.090	0.99	0.68	0.79	3.359
	313	1.80	2.390	0.071	0.093	0.87	0.66	0.75	3.596
0.10	298	1.83	2.298	0.052	0.066	1.12	0.89	0.80	3.208
	303	1.82	2.758	0.066	0.069	0.91	0.86	0.66	3.955
	308	1.81	3.678	0.081	0.081	0.75	0.75	0.49	5.416
	313	1.80	4.137	0.120	0.096	0.52	0.64	0.44	6.130
0.15	298	1.87	2.528	0.069	0.033	0.85	1.75	0.74	3.469
	303	1.86	3.678	0.070	0.057	0.86	1.04	0.51	5.118
	308	1.85	4.597	0.073	0.078	0.83	0.78	0.40	6.635
	313	1.84	5.517	0.075	0.111	0.82	0.56	0.33	8.173

Table (5-8) : *Corrosion parameters for the polarization of 7075 alloy in aerated NaOH solution pH=11 with presence of CH_3COONa at four temperatures.*

Conc. Of CH_3COONa mol.dm^{-3}	T (K)	Corrosion		b (V.decade^{-1})		α		$R_p/10^5$ ($\Omega.cm^{-2}$)	$i_c/10^{-7}$ ($A.cm^{-2}$)
		$-E_{corr}$ (V)	$i_{corr}/10^{-5}$ ($A.cm^{-2}$)	$-b_c$	$+b_a$	α_c	α_a		
0.05	298	1.93	1.038	0.071	0.060	0.83	0.99	1.86	1.380
	303	1.91	1.280	0.081	0.070	0.74	0.86	1.49	1.752
	308	1.87	1.553	0.092	0.075	0.66	0.81	1.20	2.212
	313	1.86	1.730	0.100	0.082	0.62	0.76	0.08	2.497
0.10	298	1.90	1.557	0.060	0.065	0.99	0.91	1.22	2.104
	303	1.88	2.491	0.068	0.120	0.88	0.50	0.75	3.480
	308	1.87	3.114	0.082	0.142	0.74	0.43	0.60	4.423
	313	1.86	3.806	0.115	0.176	0.54	0.35	0.49	5.504
0.15	298	1.88	2.422	0.058	0.085	1.01	0.69	0.78	3.291
	303	1.87	2.768	0.075	0.096	0.80	0.62	0.68	3.838
	308	1.86	3.541	0.081	0.136	0.75	0.44	0.53	5.008
	313	1.85	4.498	0.088	0.150	0.70	0.41	0.41	6.578

Table (5-9) : *Corrosion parameters for the polarization of pure aluminium in aerated NaOH solution pH=9 with presence of CH_3COONa at four temperatures.*

Conc. Of H_3COONa $mol.dm^{-3}$	T (K)	Corrosion		b $(V.decade^{-1})$		α		$R_p/10^7$ $(\Omega.cm^{-2})$	$i_o/10^9$ $(A.cm^{-2})$
		$-E_{corr}$ (V)	$i_{corr}/10^7$ $(A.cm^{-2})$	$-b_c$	$+b_a$	α_c	α_a		
0.05	298	1.73	0.432	0.166	0.200	0.35	0.30	4.00	0.642
	303	1.71	0.493	0.200	0.230	0.30	0.26	3.47	0.752
	308	1.69	0.586	0.230	0.300	0.26	0.20	2.88	0.922
	313	1.68	0.617	0.250	0.450	0.24	0.14	2.72	0.992
0.10	298	1.37	0.524	0.120	0.120	0.49	0.49	2.61	0.984
	303	1.31	0.802	0.250	0.130	0.24	0.46	1.63	1.601
	308	1.29	1.080	0.300	0.176	0.20	0.35	1.19	2.230
	313	1.27	1.234	0.375	0.272	0.17	0.23	1.03	2.619
0.15	298	1.82	0.601	0.069	0.098	0.85	0.59	3.03	0.847
	303	1.77	0.652	0.098	0.100	0.61	0.60	2.71	0.963
	308	1.75	0.813	0.130	0.130	0.47	0.47	2.15	1.234
	313	1.73	0.925	0.157	0.150	0.39	0.41	1.87	1.442

Table (5-10) : *Corrosion parameters for the polarization of 2024 alloy in aerated NaOH solution pH=9 with presence of CH_3COONa at four temperatures.*

Conc. Of CH_3COONa $mol.dm^{-3}$	T (K)	Corrosion		b $(V.decade^{-1})$		α		$R_p/10^7$ $(\Omega.cm^{-2})$	$i_o/10^9$ $(A.cm^{-2})$
		$-E_{corr}$ (V)	$i_{corr}/10^7$ $(A.cm^{-2})$	$-b_c$	$+b_a$	α_c	α_a		
0.05	298	1.18	0.541	0.150	0.120	0.39	0.49	2.18	1.178
	303	1.12	0.591	0.200	0.210	0.30	0.29	1.90	1.374
	308	1.09	0.689	0.331	0.300	0.18	0.20	1.58	1.680
	313	1.06	0.837	0.375	0.333	0.16	0.18	1.27	2.124
0.10	298	1.16	0.985	0.071	0.120	0.83	0.49	1.18	2.175
	303	1.14	1.115	0.075	0.142	0.80	0.42	1.02	2.559
	308	1.12	1.351	0.076	0.375	0.79	0.16	0.83	3.198
	313	1.11	1.463	0.088	0.400	0.70	0.15	0.76	3.549
0.15	298	1.20	0.620	0.300	0.075	0.20	0.78	1.94	1.323
	303	1.16	0.689	0.491	0.096	0.12	0.62	1.68	1.554
	308	1.12	0.825	0.520	0.157	0.11	0.39	1.36	1.951
	313	1.09	0.945	0.650	0.333	0.09	0.19	1.15	2.345

Table (5-11) : *Corrosion parameters for the polarization of 5083 alloy in aerated NaOH solution pH=9 with presence of CH₃COONa at four temperatures.*

Conc. Of CH₃COONa mol.dm⁻³	T (K)	Corrosion		B (V.decade⁻¹)		α		$R_p/10^7$ (Ω.cm⁻²)	$i_o/10^{-9}$ (A.cm⁻²)
		$-E_{corr}$ (V)	$i_{corr}/10^{-7}$ (A.cm⁻²)	$-b_c$	$+b_a$	α_c	α_a		
0.05	298	1.59	0.597	0.250	0.200	0.24	0.29	2.66	0.965
	303	1.55	0.689	0.300	0.214	0.20	0.28	2.25	1.160
	308	1.53	0.735	0.315	0.300	0.19	0.20	2.08	1.276
	313	1.51	0.857	0.333	0.331	0.18	0.18	1.76	1.532
0.10	298	1.26	1.517	0.075	0.200	0.78	0.29	0.83	3.093
	303	1.25	1.655	0.088	0.214	0.68	0.28	0.76	3.434
	308	1.24	1.839	0.090	0.230	0.67	0.26	0.67	3.961
	313	1.23	2.206	0.100	0.272	0.62	0.22	0.56	4.816
0.15	298	1.55	0.643	0.085	0.078	0.68	0.75	2.41	1.065
	303	1.49	0.735	0.091	0.081	0.66	0.74	2.03	1.286
	308	1.47	0.833	0.115	0.136	0.53	0.45	1.76	1.508
	313	1.35	0.965	0.157	0.187	0.39	0.33	1.40	1.927

Table (5-12) : *Corrosion parameters for the polarization of 7075 alloy in aerated NaOH solution pH=9 with presence of CH₃COONa at four temperatures.*

Conc. Of H₃COONa mol.dm⁻³	T (K)	Corrosion		b (V.decade⁻¹)		α		$R_p/10^7$ (Ω.cm⁻²)	$i_o/10^{-9}$ (A.cm⁻²)
		$-E_{corr}$ (V)	$i_{corr}/10^{-7}$ (A.cm⁻²)	$-b_c$	$+b_a$	α_c	α_a		
0.05	298	1.60	0.579	0.150	0.151	0.39	0.39	2.76	0.930
	303	1.56	0.657	0.181	0.166	0.33	0.36	2.37	1.101
	308	1.54	0.795	0.200	0.187	0.31	0.32	1.94	1.368
	313	1.52	0.849	0.250	0.200	0.25	0.31	1.79	1.507
0.10	298	1.36	1.038	0.120	0.200	0.49	0.29	1.31	1.960
	303	1.30	1.127	0.150	0.210	0.40	0.28	1.15	2.270
	308	1.28	1.484	0.142	0.230	0.42	0.26	0.86	3.086
	313	1.26	1.730	0.166	0.333	0.37	0.19	0.73	3.695
0.15	298	1.79	0.632	0.053	0.076	1.11	0.77	2.83	0.907
	303	1.74	0.692	0.082	0.088	0.73	0.68	2.51	1.040
	308	1.72	0.831	0.150	0.111	0.41	0.55	2.07	1.282
	313	1.64	0.955	0.214	0.300	0.26	0.21	1.72	1.568

5-2-2 Addition Of Sodium Chromate

The addition of sodium chromate (Na_2CrO_4) as an inorganic inhibitor to 0.1 mol.dm^{-3} NaOH solution (pH=13) remains the polarization curve shape without change as shown in Fig. (5-5) which demonstrates the polarization curve for 2024 alloy in pH=13 and in the presence of 5x10^{-3} mol.dm^{-3} sodium chromate at 298K, which is similar to that obtained for pure Al and other alloys under the same condition and in the presence of higher concentration (1x10^{-2}, and 5x10^{-2} mol.dm^{-3}) of sodium chromate at four temperature in the range (298 – 313)K.

In pH=11, (1x10^{-3} mol.dm^{-3} NaOH solution), the presence of 0.05 mol.dm^{-3} Na_2CrO_4 clearly changes the polarization curve, Fig. (5-6) show this change for pure Al where the curve contains the cathodic region (abc) section and anodic dissolution of metal which begins at (c) point and continues along (cd).

The polarization curve of 2024 alloy in pH = 11 and in the presence of 0.05 mol.dm^{-3} Na_2CrO_4 is shown in Fig. (5-7) at four different temperature. This curve includes the following :

a- The cathodic section (abc).

b- The anodic dissolution section (cd) for metal, the point (d) may correspond to "Flade potential" and "Flade current density".

c- Protection film or passive layer represented in the section (de).

d- The passive layer breakdown begins at point (f) and continues along (fg).

Figures (5-8) and (5-9) show the polarization curve of 5083 and 7075 alloy respectively in the (pH=11 + 0.05 mol.dm^{-3}) system at four temperature. The curves consists of the cathodic and anodic sections which are quite sharp and clear.

The polarization curves in the presence of 5×10^{-3} mol.dm^{-3} Na$_2$CrO$_4$ in pH=11 are generally similar to that obtained for pure Al and its alloys in the presence of 1×10^{-2} mol.dm^{-3} Na$_2$CrO$_4$. The pattern of the curve is presented in Fig. (5-10) which is consists of the cathodic and anodic Tafel regions.

The presence of (5×10^{-3} and 1×10^{-2} mol.dm^{-3} Na$_2$CrO$_4$) in pH=9 gives the pattern of curve as presented in Fig. (5-11) for pure Al and its alloys at four temperature. The cathodic region (abc) is clear and smooth follow beyond point (c), by the anodic region (cd) which was relatively short on the potential axis. The anodic region changed beyond point (d) into a stable passive region (def). The breakdown of passivity begins at point (f) and continues along (fg).

In the presence of high concentration of sodium chromate (5×10^{-2} mol.dm^{-3}), the cathodic and anodic regions become clearer and sharper and shifts toward the lower values of corrosion current density, in addition of maintaining of passive layer as shown in Fig. (5-12) for pure Al which is similar to that obtained for its alloys.

Tables (5-13) to (5-24) present data which have been derived from the appropriate polarization curves.

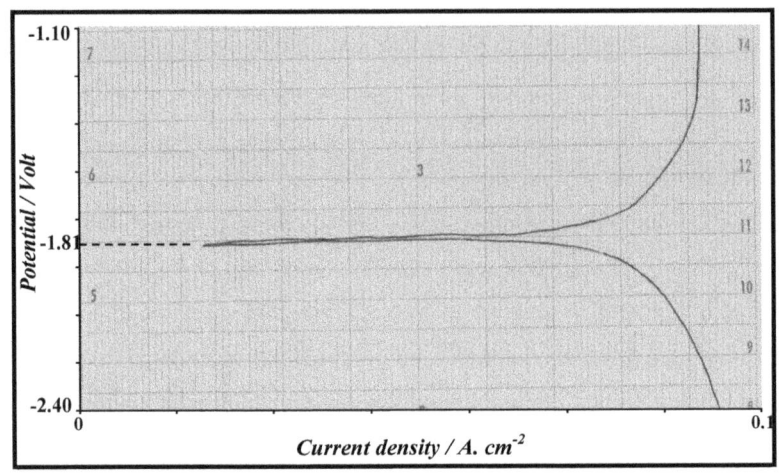

Fig. (5-5) : *The polarization curve 2024 alloy in pH=13*
in presence of 5x10^{-3} mol.dm^{-3} Na$_2$CrO$_4$ at 298K.

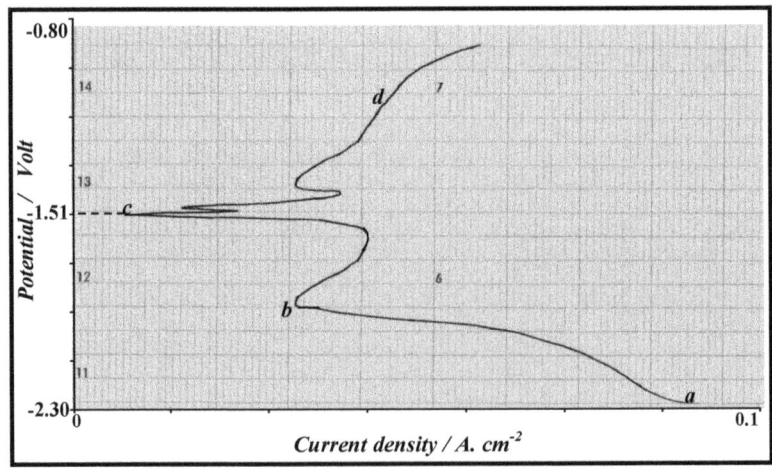

Fig. (5-6) : *The polarization curve of pure Al in pH=11*
in presence of 0.05 mol.dm^{-3} Na$_2$CrO$_4$ at 298K.

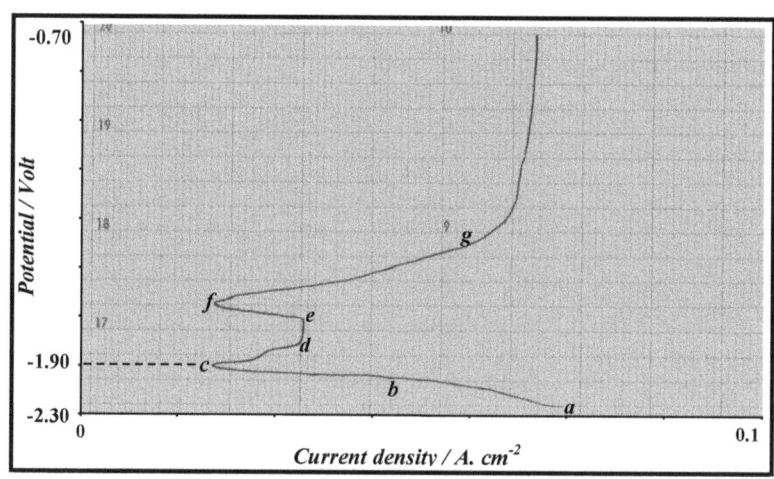

Fig. (5-7) : *The polarization curve of 2024 alloy in pH=11 in presence of 0.05 mol.dm^{-3} Na$_2$CrO$_4$ at 298K.*

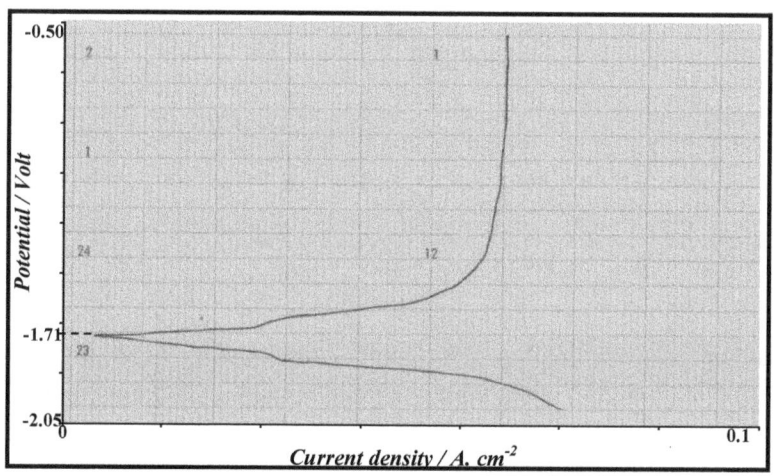

Fig. (5-8) : *The polarization curve of 5083 alloy in pH=11 in presence of 0.05 mol.dm^{-3} Na$_2$CrO$_4$ at 298K.*

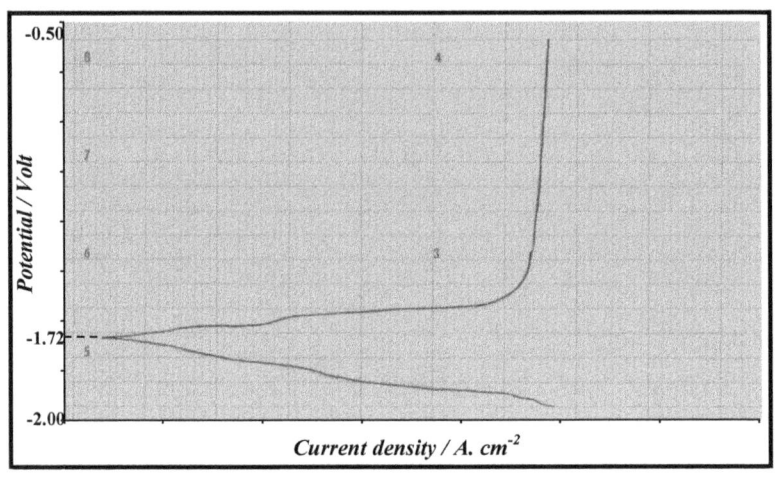

Fig. (5-9) : *The polarization curve of 7075 alloy in pH=11*
in presence of 0.05 mol.dm^{-3} Na$_2$CrO$_4$ at 298K.

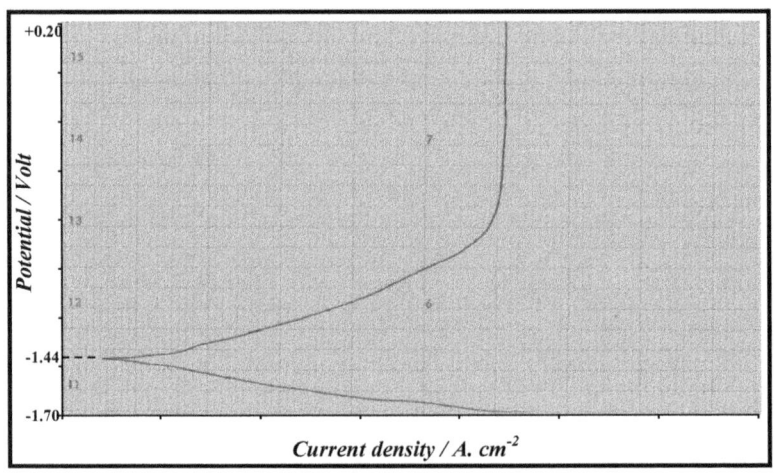

Fig. (5-10) : *The polarization curve of 5083 alloy in pH=11*
in presence of 1x10^{-2} mol.dm^{-3} Na$_2$CrO$_4$ at 298K.

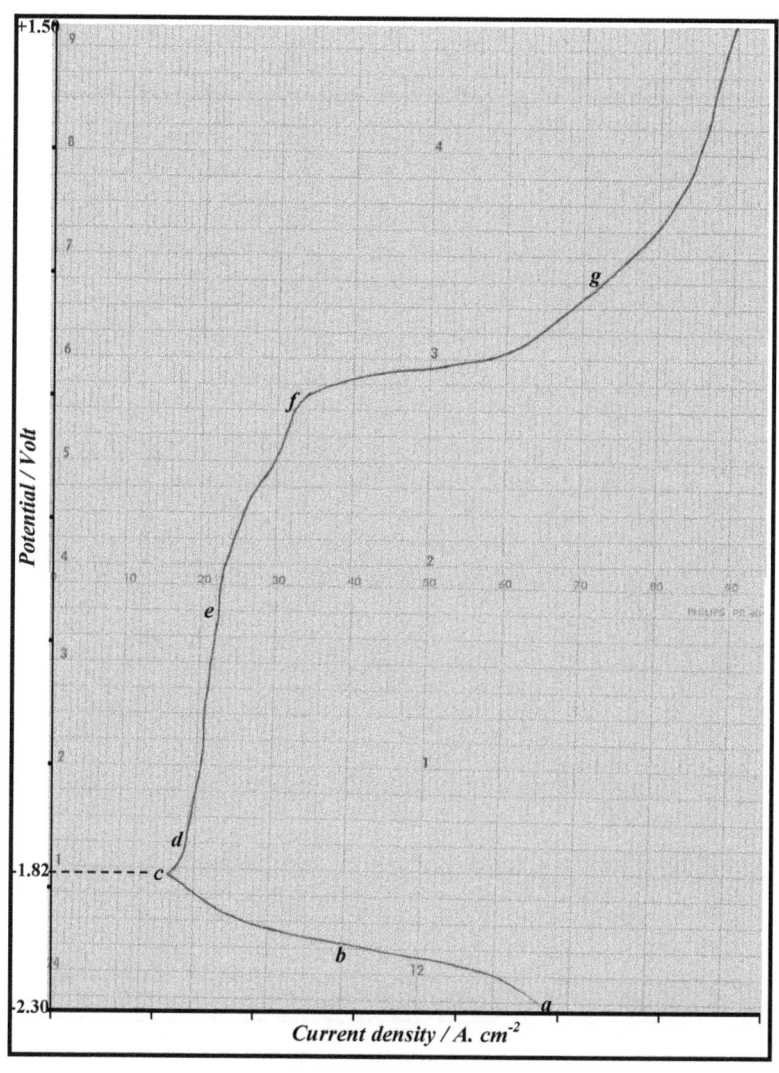

Fig. (5-11) : *The polarization curve of 5083 alloy in pH=9*
in presence of 5x10^{-3} mol.dm^{-3} Na$_2$CrO$_4$ at 298K.

145

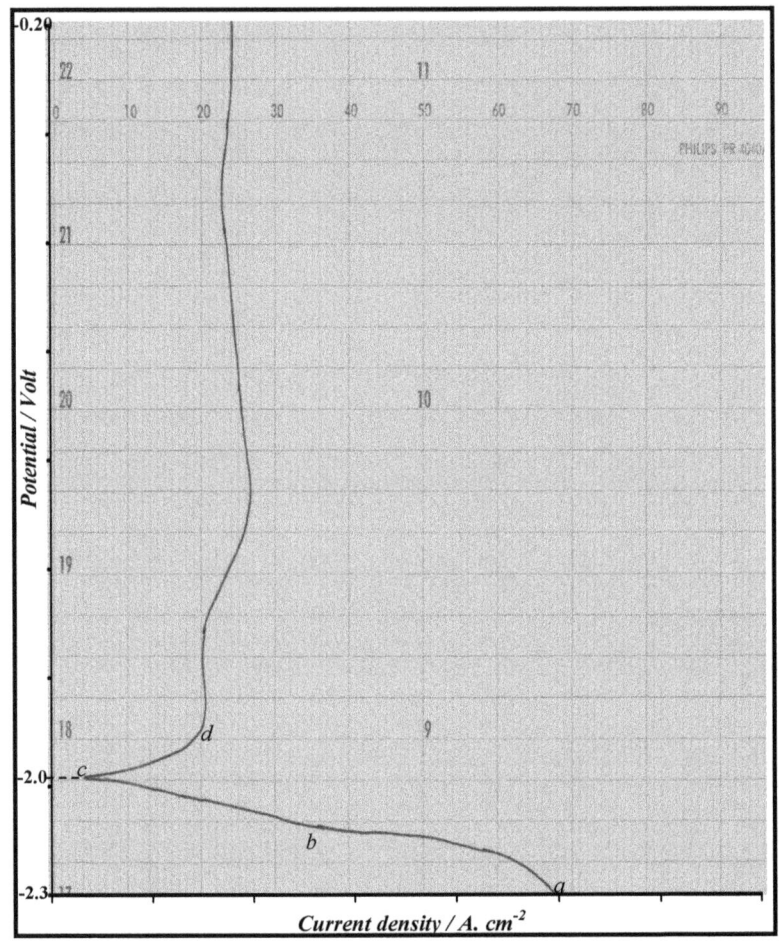

Fig. (5-12) : *The polarization curve of pure Al in pH=9*
in presence of (5x10⁻² mol.dm⁻³) Na₂CrO₄ at 298K.

Table (5-13) : *Corrosion parameters for the polarization of pure aluminium in pH=13 with presence of Na_2CrO_4 at four temperatures.*

Conc. Of Na_2CrO_4 mol.dm^{-3}	T (K)	Corrosion		b (V.decade^{-1})		α		$R_p/10^3$ ($\Omega.cm^{-2}$)	$i_o/10^{-5}$ (A.cm^{-2})
		$-E_{corr}$ (V)	$i_{corr}/10^{-3}$ (A.cm^{-2})	$-b_c$	$+b_a$	α_c	α_a		
$5x10^{-3}$	298	2.10	0.572	0.078	0.062	0.75	0.95	3.67	0.699
	303	2.09	0.630	0.083	0.071	0.72	0.84	3.32	0.786
	308	2.08	0.757	0.103	0.103	0.59	0.59	2.75	0.965
	313	2.07	0.851	0.120	0.107	0.52	0.58	2.43	1.110
$1x10^{-2}$	298	1.73	0.722	0.085	0.093	0.69	0.63	2.40	1.070
	303	1.72	0.937	0.093	0.115	0.64	0.52	1.84	1.418
	308	1.71	1.005	0.100	0.120	0.61	0.51	1.70	1.561
	313	1.70	1.117	0.115	0.130	0.53	0.47	1.52	1.774
$5x10^{-2}$	298	1.69	0.818	0.085	0.071	0.69	0.83	2.07	1.240
	303	1.68	0.980	0.088	0.093	0.68	0.64	1.71	1.526
	308	1.67	1.055	0.090	0.100	0.67	0.61	1.58	1.680
	313	1.66	1.176	0.100	0.130	0.62	0.48	1.41	1.913

Table (5-14) : *Corrosion parameters for the polarization of 2024 alloy in pH=13 with presence of Na_2CrO_4 at four temperatures.*

Conc. Of Na_2CrO_4 mol.dm^{-3}	T (K)	Corrosion		B (V.decade^{-1})		α		$R_p/10^3$ ($\Omega.cm^{-2}$)	$i_o/10^{-5}$ (A.cm^{-2})
		$-E_{corr}$ (V)	$i_{corr}/10^{-3}$ (A.cm^{-2})	$-b_c$	$+b_a$	α_c	α_a		
$5x10^{-3}$	298	1.81	0.597	0.066	0.058	0.089	1.01	3.03	0.847
	303	1.80	0.689	0.081	0.065	0.74	0.92	2.61	1.000
	308	1.79	0.791	0.088	0.073	0.69	0.83	2.26	1.174
	313	1.78	0.893	0.100	0.085	0.62	0.73	1.99	1.355
$1x10^{-2}$	298	1.48	0.873	0.075	0.085	0.78	0.69	1.70	1.510
	303	1.46	0.949	0.085	0.088	0.70	0.68	1.54	1.695
	308	1.45	1.015	0.093	0.090	0.65	0.67	1.43	1.856
	313	1.43	1.121	0.100	0.100	0.62	0.62	1.28	2.107
$5x10^{-2}$	298	1.44	0.965	0.088	0.088	0.67	0.67	1.49	1.723
	303	1.43	1.011	0.090	0.096	0.66	0.66	1.41	1.851
	308	1.42	1.149	0.093	0.120	0.65	0.51	1.24	2.140
	313	1.39	1.195	0.100	0.125	0.62	0.49	1.16	2.325

Table (5-15) : _Corrosion parameters for the polarization of 5083 alloy in_ _pH=13 with presence of Na_2CrO_4 at four temperatures._

Conc. Of Na_2CrO_4 mol.dm^{-3}	T (K)	Corrosion		b (V.decade^{-1})		α		$R_p/10^3$ (Ω.cm^{-2})	$i_c/10^{-5}$ (A.cm^{-2})
		$-E_{corr}$ (V)	$i_{corr}/10^{-3}$ (A.cm^{-2})	$-b_c$	$+b_a$	$α_c$	$α_a$		
5x10^{-3}	298	2.01	0.550	0.078	0.083	0.75	0.71	3.65	0.703
	303	2.00	0.625	0.083	0.093	0.72	0.64	3.20	0.816
	308	1.99	0.669	0.090	0.100	0.67	0.61	2.97	0.894
	313	1.98	0.714	0.100	0.136	0.62	0.45	2.77	0.974
1x10^{-2}	298	1.66	0.714	0.055	0.062	1.07	0.95	2.32	1.106
	303	1.65	0.758	0.076	0.079	0.79	0.76	2.18	1.197
	308	1.64	0.803	0.093	0.096	0.65	0.63	2.04	1.301
	313	1.63	0.848	0.100	0.100	0.62	0.62	1.92	1.405
5x10^{-2}	298	1.61	0.758	0.075	0.083	0.78	0.71	2.12	1.211
	303	1.60	0.803	0.085	0.095	0.71	0.63	1.99	1.312
	308	1.59	0.892	0.093	0.103	0.65	0.59	1.78	1.491
	313	1.58	0.982	0.100	0.115	0.62	0.53	1.61	1.675

Table (5-16) : _Corrosion parameters for the polarization of 7075 alloy in_ _pH=13 with presence of Na_2CrO_4 at four temperatures._

Conc. Of Na_2CrO_4 mol.dm^{-3}	T (K)	Corrosion		b (V.decade^{-1})		α		$R_p/10^3$ (Ω.cm^{-2})	$i_c/10^{-5}$ (A.cm^{-2})
		$-E_{corr}$ (V)	$i_{corr}/10^{-3}$ (A.cm^{-2})	$-b_c$	$+b_a$	$α_c$	$α_a$		
5x10^{-3}	298	2.00	0.336	0.052	0.066	1.13	0.89	5.95	0.431
	303	1.99	0.403	0.069	0.083	0.87	0.72	4.94	0.528
	308	1.98	0.470	0.075	0.093	0.81	0.65	4.21	0.630
	313	1.97	0.504	0.100	0.136	0.62	0.45	3.91	0.690
1x10^{-2}	298	1.64	0.208	0.037	0.040	1.59	1.47	7.88	0.326
	303	1.61	0.268	0.056	0.069	1.07	0.87	6.01	0.434
	308	1.58	0.369	0.066	0.082	0.92	0.74	4.28	0.620
	313	1.56	0.436	0.078	0.100	0.79	0.62	3.58	0.753
5x10^{-2}	298	1.57	0.194	0.066	0.071	0.89	0.83	8.09	0.317
	303	1.56	0.258	0.071	0.083	0.84	0.72	6.05	0.431
	308	1.55	0.285	0.088	0.097	0.69	0.62	5.44	0.488
	313	1.54	0.336	0.095	0.107	0.65	0.58	4.58	0.589

Table (5-17) : *Corrosion parameters for the polarization of pure aluminium in pH=11 with presence of Na_2CrO_4 at four temperatures.*

Conc. Of Na_2CrO_4 mol.dm^{-3}	T (K)	Corrosion		b (V.decade^{-1})		α		$R_p/10^7$ ($\Omega.cm^{-2}$)	$i_c/10^{-9}$ ($A.cm^{-2}$)
		$-E_{corr}$ (V)	$i_{corr}/10^{-7}$ ($A.cm^{-2}$)	$-b_c$	$+b_a$	$α_c$	$α_a$		
5x10^{-3}	298	2.00	0.522	0.075	0.107	0.78	0.55	3.83	0.670
	303	1.99	0.845	0.111	0.125	0.54	0.48	2.36	1.106
	308	1.96	0.970	0.120	0.157	0.51	0.38	2.02	1.314
	313	1.94	1.133	0.136	0.200	0.45	0.31	1.71	1.577
1x10^{-2}	298	1.52	0.915	0.111	0.115	0.53	0.51	1.66	1.546
	303	1.49	1.547	0.120	0.135	0.50	0.44	0.96	2.719
	308	1.47	1.961	0.166	0.150	0.38	0.41	0.75	3.539
	313	1.45	2.535	0.176	0.187	0.35	0.33	0.57	4.732
5x10^{-2}	298	1.51	0.816	0.026	0.040	2.27	1.47	1.85	1.388
	303	1.47	0.949	0.034	0.056	1.76	1.07	1.55	1.684
	308	1.44	1.015	0.093	0.100	0.65	0.61	1.42	1.869
	313	1.42	1.193	0.150	0.120	0.41	0.52	1.19	2.266

Table (5-18) : *Corrosion parameters for the polarization of 2024 alloy in pH=11 with presence of Na_2CrO_4 at four temperatures.*

Conc. Of Na_2CrO_4 mol.dm^{-3}	T (K)	Corrosion		b (V.decade^{-1})		α		$R_p/10^7$ ($\Omega.cm^{-2}$)	$i_c/10^{-9}$ ($A.cm^{-2}$)
		$-E_{corr}$ (V)	$i_{corr}/10^{-7}$ ($A.cm^{-2}$)	$-b_c$	$+b_a$	$α_c$	$α_a$		
5x10^{-3}	298	1.84	1.149	0.050	0.069	1.18	0.85	1.60	1.604
	303	1.83	1.333	0.083	0.100	0.72	0.60	1.37	1.905
	308	1.82	1.379	0.100	0.103	0.61	0.59	1.32	2.011
	313	1.81	1.701	0.120	0.120	0.52	0.52	1.06	2.544
1x10^{-2}	298	1.46	3.586	0.046	0.073	1.28	0.81	0.41	6.261
	303	1.45	3.770	0.050	0.075	1.20	0.80	0.38	6.868
	308	1.43	4.137	0.058	0.120	1.05	0.50	0.35	7.583
	313	1.42	4.597	0.060	0.157	1.03	0.39	0.31	8.700
5x10^{-2}	298	1.90	2.988	0.030	-	1.97	-	0.64	4.011
	303	1.87	3.218	0.037	-	1.62	-	0.58	4.500
	308	1.84	3.448	0.045	-	1.35	-	0.53	5.008
	313	1.83	3.678	0.078	-	0.79	-	0.50	5.394

Table (5-19) : *Corrosion parameters for the polarization of 5083 alloy in pH=11 with presence of Na_2CrO_4 at four temperatures.*

Conc. Of Na_2CrO_4 mol.dm^{-3}	T (K)	Corrosion		b (V.decade^{-1})		a		$R_p/10^7$ ($\Omega.cm^{-2}$)	$i_o/10^{-9}$ ($A.cm^{-2}$)
		$-E_{corr}$ (V)	$i_{corr}/10^{-7}$ ($A.cm^{-2}$)	$-b_c$	$+b_a$	a_c	a_a		
5x10^{-3}	298	1.87	0.892	0.058	0.075	1.11	1.11	2.10	1.222
	303	1.86	0.937	0.060	0.100	1.00	0.60	1.99	1.312
	308	1.85	0.982	0.071	0.115	0.86	0.53	1.88	1.412
	313	1.84	1.071	0.157	0.200	0.39	0.31	1.72	1.568
1x10^{-2}	298	1.44	2.008	0.085	0.130	0.69	0.45	0.72	3.565
	303	1.42	2.232	0.107	0.145	0.56	0.41	0.64	4.078
	308	1.41	2.678	0.120	0.166	0.51	0.36	0.53	5.008
	313	1.40	3.437	0.150	0.187	0.41	0.33	0.41	6.578
5x10^{-2}	298	1.71	1.473	0.044	0.050	1.34	1.18	1.16	2.213
	303	1.70	1.562	0.046	0.055	1.31	1.09	1.09	2.394
	308	1.65	1.785	0.050	0.061	1.22	1.00	0.92	2.885
	313	1.60	1.919	0.055	0.070	1.12	0.88	0.83	3.250

Table (5-20) : *Corrosion parameters for the polarization of 7075 alloy in pH=11 with presence of Na_2CrO_4 at four temperatures.*

Conc. Of Na_2CrO_4 mol.dm^{-3}	T (K)	Corrosion		b (V.decade^{-1})		a		$R_p/10^7$ ($\Omega.cm^{-2}$)	$i_o/10^{-9}$ ($A.cm^{-2}$)
		$-E_{corr}$ (V)	$i_{corr}/10^{-7}$ ($A.cm^{-2}$)	$-b_c$	$+b_a$	a_c	a_a		
5x10^{-3}	298	1.97	0.336	0.037	0.071	1.59	0.83	5.86	0.438
	303	1.95	0.436	0.062	0.088	0.96	0.68	4.47	0.584
	308	1.93	0.470	0.073	0.093	0.83	0.65	4.11	0.646
	313	1.92	0.537	0.085	0.107	0.73	0.58	3.58	0.753
1x10^{-2}	298	1.42	0.773	0.100	0.048	0.59	1.23	1.84	1.395
	303	1.41	0.840	0.120	0.053	0.50	1.13	1.68	1.554
	308	1.40	1.008	0.125	0.081	0.48	0.75	1.39	1.909
	313	1.39	1.176	0.142	0.120	0.43	0.51	1.18	2.286
5x10^{-2}	298	1.72	0.672	0.093	0.037	0.63	1.59	2.56	1.003
	303	1.70	0.739	0.100	0.047	0.60	1.27	2.30	1.135
	308	1.68	0.806	0.120	0.050	0.50	1.22	2.08	1.276
	313	1.64	0.840	0.125	0.063	0.49	0.98	1.95	1.383

Table (5-21) : *Corrosion parameters for the polarization of pure aluminium in pH=9 with presence of Na_2CrO_4 at four temperatures.*

Conc. Of Na_2CrO_4 mol.dm^{-3}	T (K)	Corrosion		b (V.decade^{-1})		α		$R_p/10^7$ ($\Omega.cm^{-2}$)	$i_o/10^{-9}$ ($A.cm^{-2}$)
		$-E_{corr}$ (V)	$i_{corr}/10^{-7}$ ($A.cm^{-2}$)	$-b_c$	$+b_a$	α_c	α_a		
$5x10^{-3}$	298	1.87	4.048	0.200	0.233	0.29	0.25	0.46	5.580
	303	1.85	4.475	0.231	0.240	0.26	0.25	0.41	6.366
	308	1.84	4.901	0.250	0.265	0.24	0.23	0.38	6.984
	313	1.82	5.228	0.272	0.271	0.22	0.22	0.35	7.706
$1x10^{-2}$	298	1.54	1.960	0.157	0.210	0.37	0.28	0.79	3.249
	303	1.52	2.287	0.166	0.230	0.36	0.26	0.66	3.955
	308	1.51	3.267	0.200	0.241	0.30	0.25	0.46	5.770
	313	1.50	4.048	0.230	0.250	0.26	0.24	0.37	7.289
$5x10^{-2}$	298	2.00	0.633	0.071	0.051	0.83	1.15	3.16	0.812
	303	1.98	0.648	0.095	0.060	0.63	1.00	3.06	0.853
	308	1.96	0.716	0.120	0.071	0.50	0.86	2.74	0.969
	313	1.94	0.845	0.130	0.085	0.47	0.73	2.30	1.173

Table (5-22) : *Corrosion parameters for the polarization of 2024 alloy in pH=9 with presence of Na_2CrO_4 at four temperatures.*

Conc. Of Na_2CrO_4 mol.dm^{-3}	T (K)	Corrosion		b (V.decade^{-1})		α		$R_p/10^7$ ($\Omega.cm^{-2}$)	$i_o/10^{-9}$ ($A.cm^{-2}$)
		$-E_{corr}$ (V)	$i_{corr}/10^{-7}$ ($A.cm^{-2}$)	$-b_c$	$+b_a$	α_c	α_a		
$5x10^{-3}$	298	1.98	4.137	0.060	0.056	0.98	1.05	0.48	5.348
	303	1.91	5.057	0.068	0.060	0.88	1.00	0.38	6.868
	308	1.86	5.977	0.103	0.069	0.59	0.88	0.31	8.561
	313	1.84	7.356	0.107	0.120	0.58	0.52	0.25	10.788
$1x10^{-2}$	298	1.30	3.578	0.044	0.058	1.34	1.01	0.36	7.131
	303	1.25	4.137	0.050	0.062	1.20	0.96	0.30	8.700
	308	1.23	4.597	0.054	0.078	1.13	0.78	0.27	9.830
	313	1.21	4.857	0.060	0.115	1.03	0.53	0.25	10.788
$5x10^{-2}$	298	1.75	0.697	0.120	0.120	0.49	0.49	2.51	1.023
	303	1.67	0.735	0.125	0.125	0.48	0.48	2.27	1.150
	308	1.63	0.829	0.166	0.150	0.36	0.41	1.97	1.347
	313	1.62	1.075	0.230	0.187	0.26	0.33	1.51	1.786

Table (5-23) : *Corrosion parameters for the polarization of 5083 alloy in pH=9 with presence of Na_2CrO_4 at four temperatures.*

Conc. Of Na_2CrO_4 mol.dm^{-3}	T (K)	Corrosion		b (V.decade^{-1})		α		$R_p/10^7$ (Ω.cm^{-2})	$i_c/10^{-9}$ (A.cm^{-2})
		$-E_{corr}$ (V)	$i_{corr}/10^{-7}$ (A.cm^{-2})	$-b_c$	$+b_a$	$α_c$	$α_a$		
5x10^{-3}	298	1.82	5.803	0.050	0.060	1.18	0.98	0.31	8.281
	303	1.81	6.250	0.066	0.061	0.91	0.98	0.29	9.000
	308	1.80	7.589	0.120	0.081	0.51	0.75	0.24	11.058
	313	1.79	8.482	0.176	0.120	0.35	0.52	0.21	12.843
1x10^{-2}	298	1.39	3.671	0.055	0.103	1.07	0.57	0.38	6.755
	303	1.38	4.241	0.063	0.100	0.95	0.60	0.33	7.909
	308	1.36	4.910	0.136	0.176	0.45	0.35	0.28	9.479
	313	1.35	5.803	0.150	0.214	0.41	0.29	0.23	11.726
5x10^{-2}	298	1.79	0.705	0.065	0.085	0.91	0.69	2.54	1.011
	303	1.76	0.758	0.068	0.092	0.88	0.65	2.32	1.125
	308	1.72	0.848	0.075	0.100	0.81	0.61	2.03	1.307
	313	1.63	1.103	0.100	0.120	0.62	0.52	1.48	1.822

Table (5-24) : *Corrosion parameters for the polarization of 7075 alloy in pH=9 with presence of Na_2CrO_4 at four temperatures.*

Conc. Of Na_2CrO_4 mol.dm^{-3}	T (K)	Corrosion		b (V.decade^{-1})		α		$R_p/10^7$ (Ω.cm^{-2})	$i_c/10^{-9}$ (A.cm^{-2})
		$-E_{corr}$ (V)	$i_{corr}/10^{-7}$ (A.cm^{-2})	$-b_c$	$+b_a$	$α_c$	$α_a$		
5x10^{-3}	298	1.89	4.113	0.120	0.142	0.49	0.42	0.46	5.580
	303	1.87	4.569	0.130	0.176	0.46	0.34	0.41	6.366
	308	1.85	5.042	0.136	0.200	0.44	0.30	0.37	7.173
	313	1.83	5.714	0.200	0.250	0.31	0.25	0.32	8.428
1x10^{-2}	298	1.40	3.025	0.120	0.115	0.49	0.51	0.46	5.580
	303	1.39	3.361	0.152	0.125	0.39	0.48	0.41	6.366
	308	1.37	3.697	0.214	0.130	0.28	0.47	0.37	7.173
	313	1.36	4.133	0.250	0.200	0.24	0.31	0.33	8.173
5x10^{-2}	298	1.93	0.640	0.111	0.115	0.53	0.51	3.02	0.850
	303	1.91	0.707	0.115	0.136	0.52	0.44	2.70	0.967
	308	1.86	0.818	0.133	0.142	0.45	0.43	2.27	1.169
	313	1.80	0.975	0.150	0.160	0.41	0.38	1.85	1.458

5-3 *Results Of Corrosion Potentials* (E_{corr})

Inhibitors may be classified as anodic, cathodic or mixed inhibitors according to the interference with the corrosion reactions by preferentially attaching themselves to anodic, cathodic areas or both[156].

Anodic inhibitors usually function in neutral or alkaline solutions and act by producing a passivating oxide film primarily at these parts of the surface where metal cations are formed at anodic sites.

Since these sites appear randomly at the surface, the whole surface becomes covered by the passive and protecting film[157, 158]. Cathodic inhibitors act by inhibiting the cathodic regions without greatly affecting the anodic sites. Alternatively, cathodic inhibitors may operate by filming the cathodic areas. Anodic and cathodic inhibitors can be distinguished experimentally by observing the effect they produce on the corrosion potential (E_{corr}) of the metal, anodic inhibitors shift the corrosion potential to more noble while cathodic inhibitors to more active. Mixed inhibitors, however, which operate by filming the metal surface, affect both the anodic and cathodic processes, thus, the corrosion potential is a little affected.

The presence of sodium acetate in the 0.1 and 1×10^{-3} mol.dm^{-3} NaOH solution (pH = 13 and 11) shift the (E_{corr}) to more negative values (active direction) as compared with the corresponding values in the absence of sodium acetate indicating an increasing tendency of the electrode for corrosion. But the presence of sodium acetate in the pH = 9, generally, shift the (E_{corr}) to the noble direction. Figs. (5-13) to (5-21) show the effect of sodium acetate on the value of the corrosion potentials of pure Al and its alloys in the NaOH solution in the presence of sodium acetate (0.05,0.10,and 0.15mol.dm^{-3}) which may be arranged from more negative to less in a sequence as :

-E$_{corr}$ in presence of CH$_3$COONa   pure Al > 7075 > 5083 > 2024
                                            _____→
                                              Noble Direction

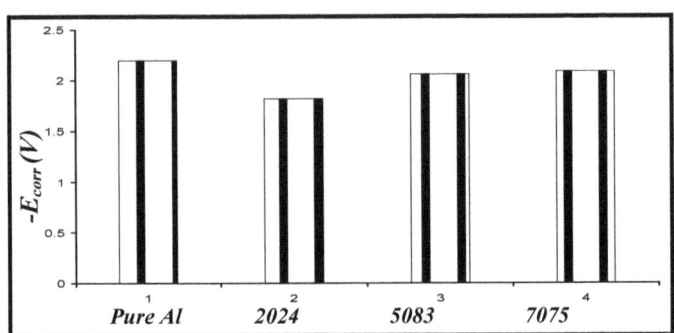

Figure (5-13) : *Values of E_{corr} plotted for pure Al and its alloys in*
pH=13 in the presence of 0.05 mol.dm^{-3} CH_3COONa at 298K.

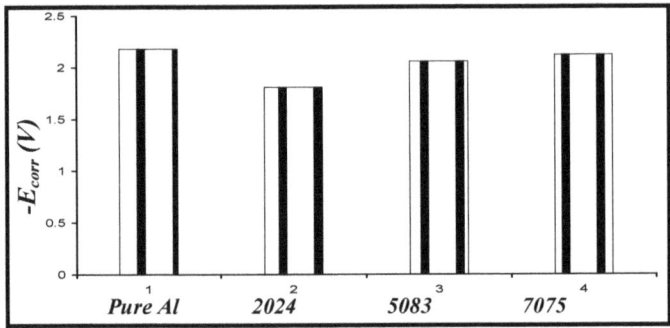

Figure (5-14) : *Values of E_{corr} plotted for pure Al and its alloys in*
pH=13 in the presence of 0.10 mol.dm^{-3} CH_3COONa at 298K.

154

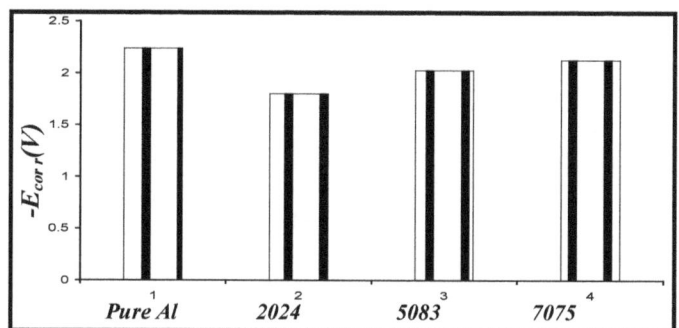

Figure (5-15) *: Values of E_{corr} plotted for pure Al and its alloys in pH=13 in the presence of 0.15 mol.dm^{-3} CH_3COONa at 298K.*

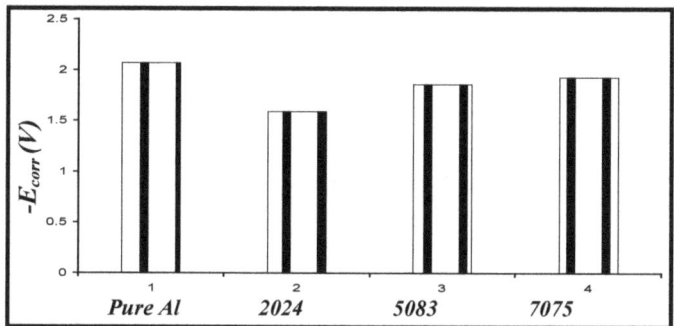

Figure (5-16) *: Values of E_{corr} plotted for pure Al and its alloys in pH=11 in the presence of 0.05 mol.dm^{-3} CH_3COONa at 298K.*

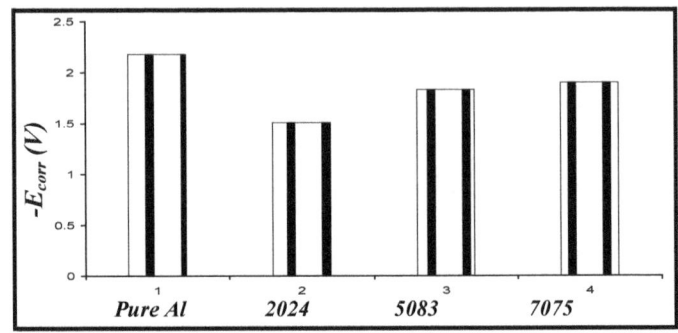

__Figure (5-17)__ : Values of E_{corr} plotted for pure Al and its alloys in pH=11 in the presence of 0.10 mol.dm^{-3} CH$_3$COONa at 298K.

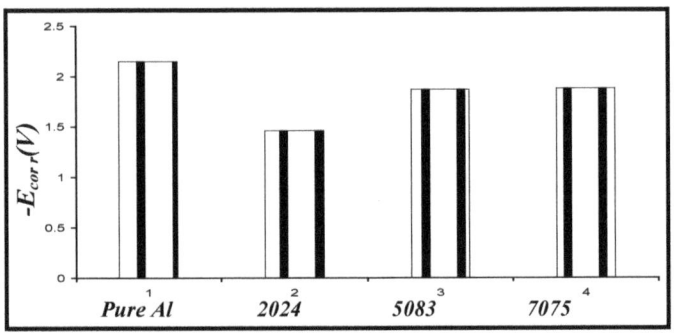

__Figure (5-18)__ : Values of E_{corr} plotted for pure Al and its alloys in pH=11 in the presence of 0.15 mol.dm^{-3} CH$_3$COONa at 298K.

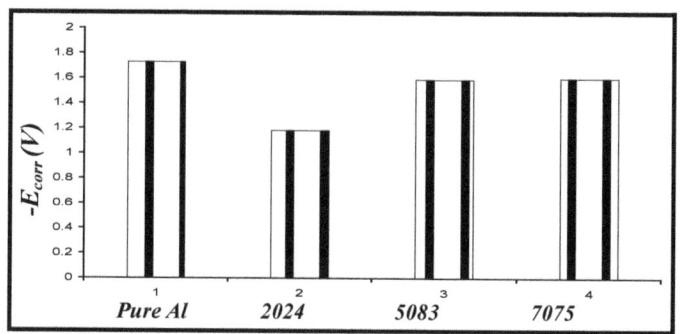

Figure (5-19) _: Values of E_{corr} plotted for pure Al and its alloys in_ $pH=9$ _in the presence of_ $0.05 \ mol.dm^{-3} \ CH_3COONa$ _at 298K._

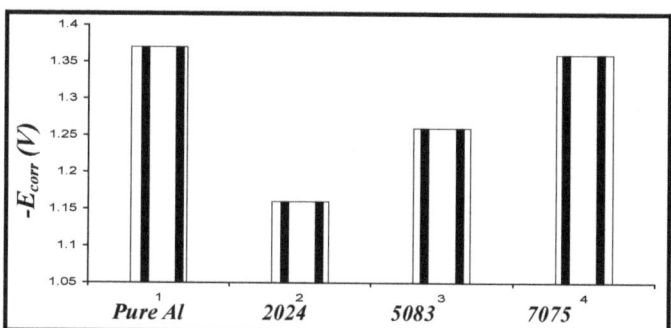

Figure (5-20) _: Values of E_{corr} plotted for pure Al and its alloys in_ $pH=9$ _in the presence of_ $0.10 \ mol.dm^{-3} \ CH_3COONa$ _at 298K._

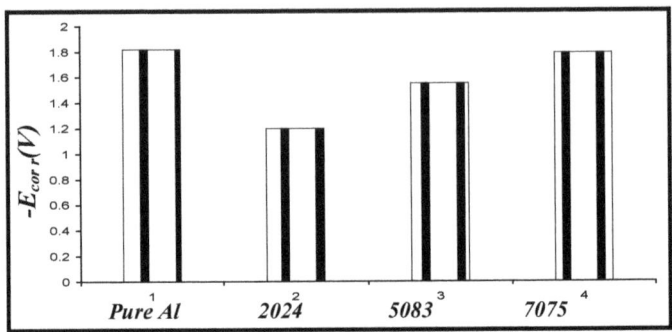

Figure (5-21) : *Values of E_{corr} plotted for pure Al and its alloys in
pH=9 in the presence of 0.15 mol.dm^{-3} CH$_3$COONa at 298K.*

The presence of sodium chromate in the 0.1 mol.dm^{-3} NaOH solution
(pH=13) shifts the (E_{corr}) to more negative values at a low concentration of the
inhibitor (5x10^{-3} mol.dm^{-3}). But the presence of a higher concentration of
inhibitors (1x10^{-2} and 5x10^{-2} mol.dm^{-3}) shifts the (E_{corr}) to noble direction in
the same pH value for pure Al and its alloys at
four temperatures.

In the lower concentration of the medium (pH=11 and 9), the presence
of (5x10^{-3} and 5x10^{-2} mol.dm^{-3}) Na$_2$CrO$_4$ shifts (E_{corr}) to more negative values,
while the presence of (1x10^{-2} mol.dm^{-3}) Na$_2$CrO$_4$ (at same medium) shifts the
(E_{corr}) toward noble values for pure Al and its alloys (2024, 5083, and 7075) at
four temperatures.

The effect of sodium chromate on the value of the corrosion potentials
of pure Al and its alloys shown in Figs. (5-22) to (5-30) at three pH values in
the presence of three concentration of inhibitor at 298K, which may be
arranged from more negative to less in a sequence as :

158

$-E_{corr}$ $(pH=13 + Na_2CrO_4)$ $pure\ Al > 5083 > 7075 > 2024$

$(pH=11 + 5x10^{-3}mol.dm^{-3}Na_2CrO_4)$ $pure\ Al>7075>5083>2024$
$-E_{corr}$ $(pH=11 +1x10^{-2}mol.dm^{-3}\ Na_2CrO_4)$ $pure\ Al>2024>5083>7075$
$(pH=11 + 5x10^{-2}mol.dm^{-3}\ Na_2CrO_4)$ $2024>7075>5083>pure\ Al$

$-E_{corr}$ $(pH=9 +5x10^{-3}mol.dm^{-3}\ Na_2CrO_4)$ $2024>7075>pure\ Al>5083$
$(1x10^{-2}\ and\ 5x10^{-2}\ mol.dm^{-3}\ Na_2CrO_4)$ $pure\ Al>7075>5083>2024$

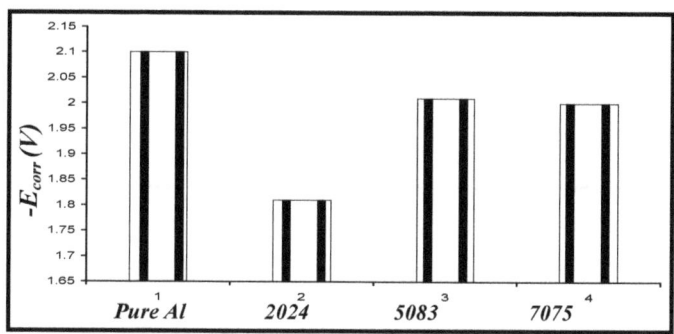

Figure (5-22) : _Values of E_{corr} plotted for pure Al and its alloys in pH=13 in the presence of 5×10^{-3} mol.dm^{-3} Na$_2$CrO$_4$ at 298 K._

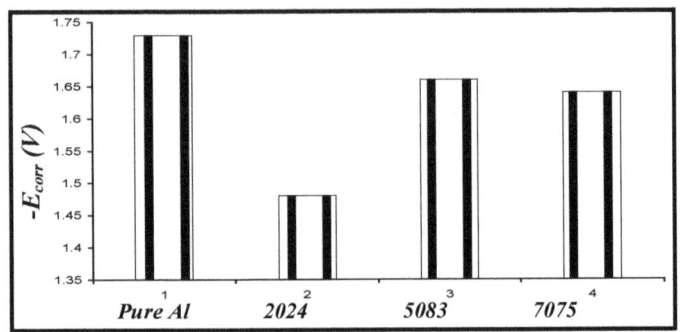

Figure (5-23) : _Values of E_{corr} plotted for pure Al and its alloys in pH=13 in the presence of 1×10^{-2} mol.dm^{-3} Na$_2$CrO$_4$ at 298 K._

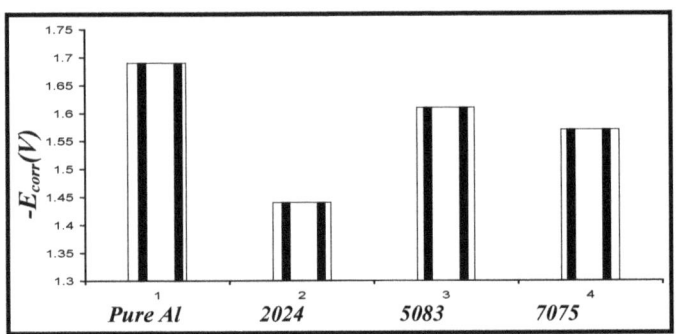

Figure (5-24) : _Values of E_{corr} plotted for pure Al and its alloys in pH=13 in the presence of 5×10^{-2} mol.dm^{-3} Na$_2$CrO$_4$ at 298 K._

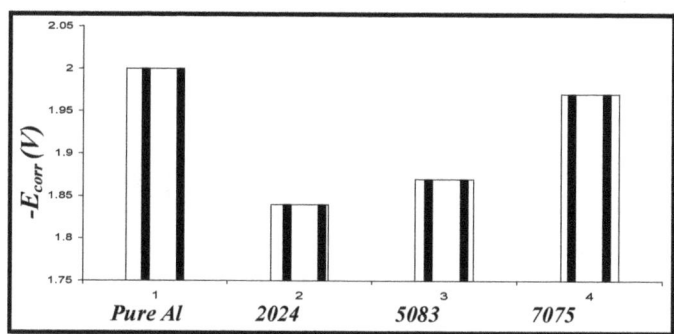

Figure (5-25) : *Values of E_{corr} plotted for pure Al and its alloys in pH=11 in the presence of 5×10^{-3} mol.dm^{-3} Na$_2$CrO$_4$ at 298 K.*

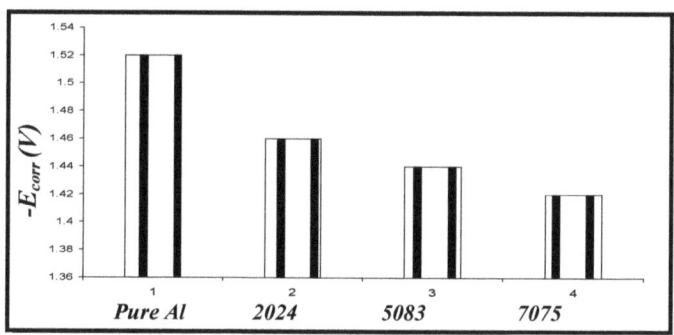

Figure (5-26) : *Values of E_{corr} plotted for pure Al and its alloys in pH=11 in the presence of 1×10^{-2} mol.dm^{-3} Na$_2$CrO$_4$ at 298 K.*

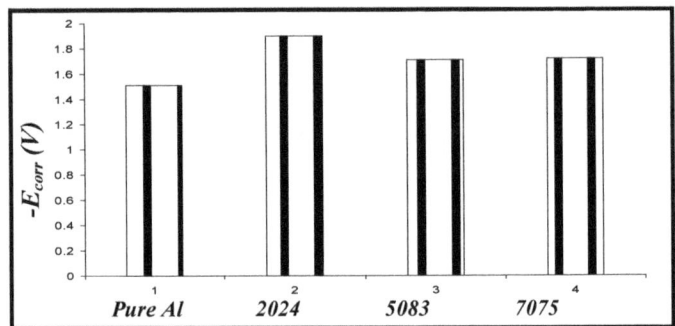

Figure (5-27) _: Values of E_{corr} plotted for pure Al and its alloys in pH=11 in the presence of 5×10^{-2} mol.dm^{-3} Na$_2$CrO$_4$ at 298 K._

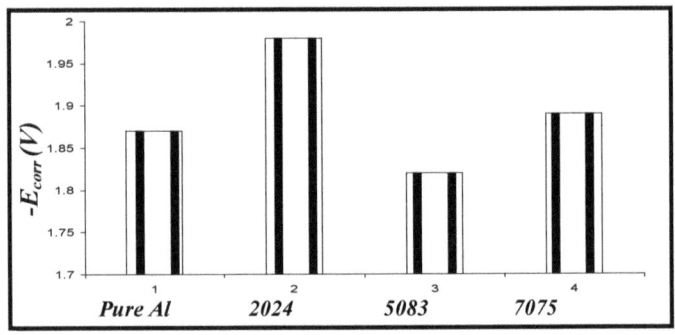

Figure (5-28) _: Values of E_{corr} plotted for pure Al and its alloys in pH=9 in the presence of 5×10^{-3} mol.dm^{-3} Na$_2$CrO$_4$ at 298 K._

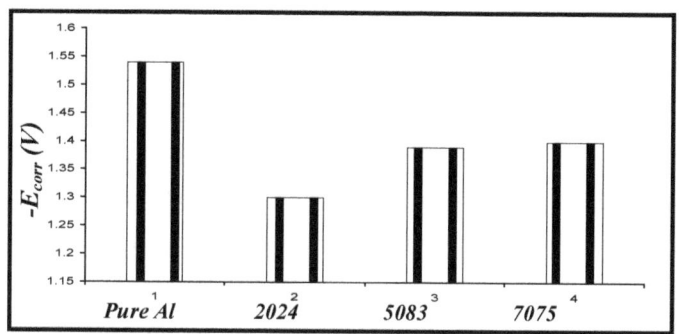

Figure (5-29) : *Values of E_{corr} plotted for pure Al and its alloys in pH=9 in the presence of 1 x 10^{-2} mol.dm^{-3} Na_2CrO_4 at 298 K.*

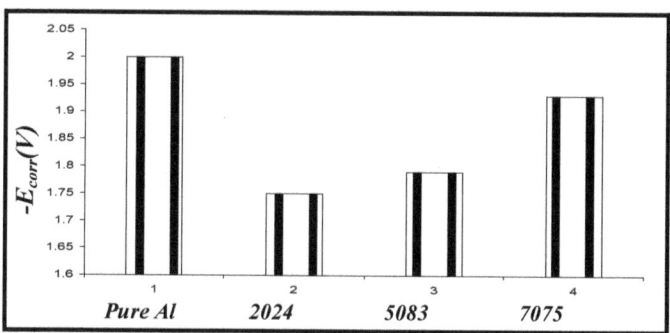

Figure (5-30) : *Values of E_{corr} plotted for pure Al and its alloys in pH=9 in the presence of 5 x 10^{-2} mol.dm^{-3} Na_2CrO_4 at 298 K.*

5-4 Results Of Corrosion Current Densities (i_{corr})

The inhibition of corrosion by chemical control of the environment is frequently defined in an electrochemical term[159], since corrosion itself is a combination of at least two electrochemical electrode reactions. It follows that

163

if the velocities of either or both of these reactions can be reduced, then some degree of inhibition of the corrosion will be achieved. So a simple definition of the inhibitor is:

It is a chemical substance, when added in small concentrations to an environment, effectively decreases the corrosion rate of the metal which is exposed to such environment[160], the corrosion current density (i_{corr}) represents the rate of corrosion under equilibrium condition.

Both of sodium acetate and sodium chromate in the used range of concentration doesn't inhibit of pure Al and its alloys in pH=13 because of high concentration of the medium (0.1 $mol.dm^{-3}$ NaOH) where :

$$(i_{corr}) \text{ with inhibitors} \quad >> \quad (i_{corr}) \text{ without inhibitors}$$

The presence of sodium acetate in pH=11 inhibits the pure Al and 2024 alloy only (doesn't inhibit the other alloys), and the corrosion current density decreases with the decreasing of the concentration of CH_3COONa.

While the presence of sodium chromate in pH=11 inhibits the pure Al and its three alloys, the corrosion current density decreases with the decreasing of the concentration of Na_2CrO_4.

In the lower concentration of NaOH solution (pH=9), sodium acetate inhibits the pure Al and its three alloys and a lower value of (i_{corr}) is at 0.05 $mol.dm^{-3}$ of CH_3COONa. Also; sodium chromate behaves the same behaviour of sodium acetate in this value of pH. But (i_{corr}) values decrease with the increasing concentration of sodium chromate (5×10^{-2} $mol.dm^{-3}$).

The results of Figs (5-31) to (5-39) and Figs. (5-40) to (5-48) show the effect of sodium acetate and sodium chromate respectively on the value of the corrosion current densities of pure aluminium and its alloys in NaOH solution at (pH=13, 11, and 9) with three concentrations of each inhibitor at 298K.

164

It is possible to conclude the following sequences :

i_{corr} *(pH=13 + CH₃COONa)*   *5083 > 2024 > pure Al > 7075*

i_{corr} *(pH=11 + CH₃COONa)*   *5083 > 7075 > pure Al > 2024*

i_{corr} *(pH=9 + CH₃COONa)*   *5083 > 7075 > 2024 > pure Al*

i_{corr} *(pH=13 + Na₂CrO₄)*   *2024 > pure Al > 5083 > 7075*

i_{corr} *(pH=11 + Na₂CrO₄)*   *2024 > 5083 > pure Al > 7075*

i_{corr} *(pH=9 + Na₂CrO₄)*   *5083 > 2024 > 7075 > pure Al*

While (i_{corr}) values for pure Al are higher than those of its alloys in the absence of additives, and the (i_{corr}) sequence for aluminium alloys is:

i_{corr} *(pH=13 and 11)*   *2024 > 5083 > 7075*

and

i_{corr} *(pH=9)*   *7075 > 5083 > 2024*

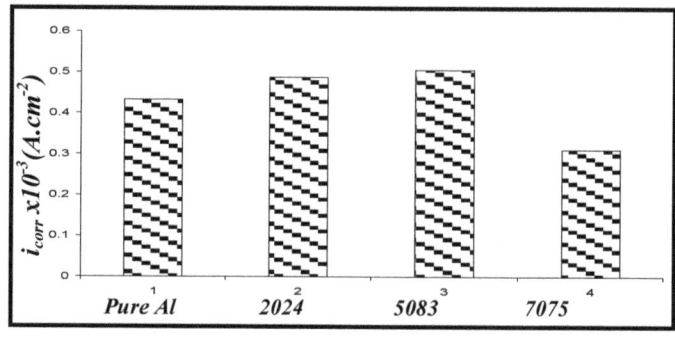

Figure (5-31) : *Values of i_{corr} plotted for pure Al and its alloys in pH=13 in the presence of 0.05 mol.dm⁻³ CH₃COONa at 298 K.*

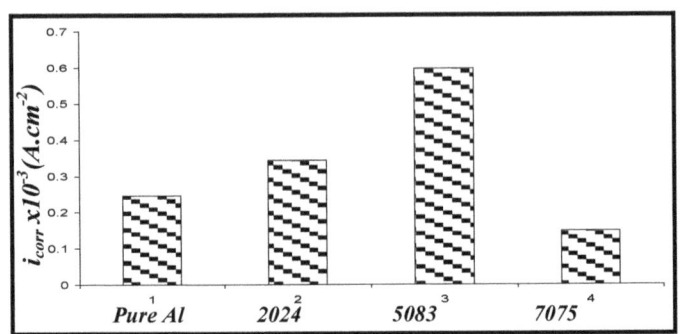

Figure (5-32) : Values of i_{corr} plotted for pure Al and its alloys in pH=13 in the presence of 0.10 mol.dm^{-3} CH$_3$COONa at 298 K.

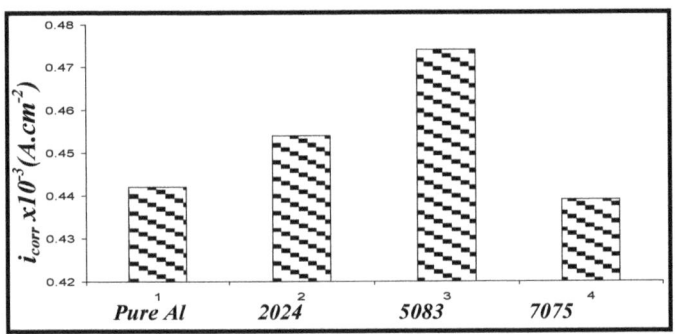

Figure (5-33) : Values of i_{corr} plotted for pure Al and its alloys in pH=13 in the presence of 0.15 mol.dm^{-3} CH$_3$COONa at 298 K.

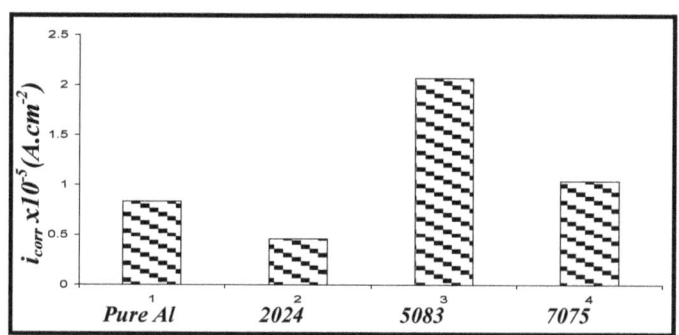

Figure (5-34) _: Values of i_{corr} plotted for pure Al and its alloys in pH=11 in the presence of 0.05 mol.dm^{-3} CH$_3$COONa at 298 K._

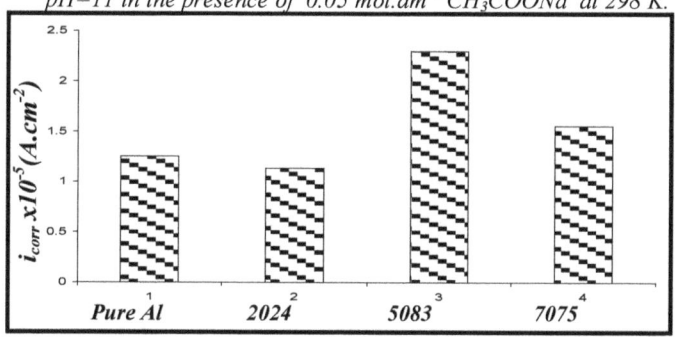

Figure (5-35) _: Values of i_{corr} plotted for pure Al and its alloys in pH=11 in the presence of 0.10 mol.dm^{-3} CH$_3$COONa at 298 K._

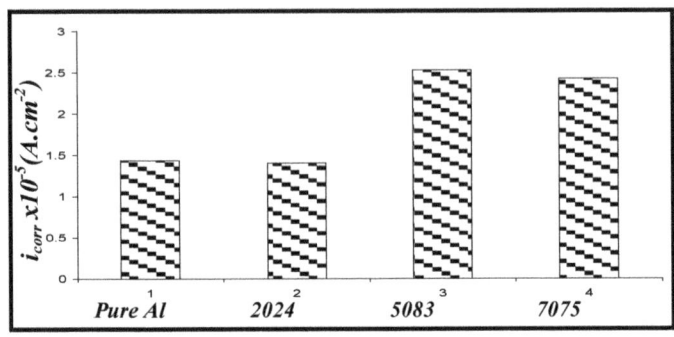

Figure (5-36) : *Values of i$_{corr}$ plotted for pure Al and its alloys in pH=11 in the presence of 0.15 mol.dm^{-3} CH$_3$COONa at 298 K.*

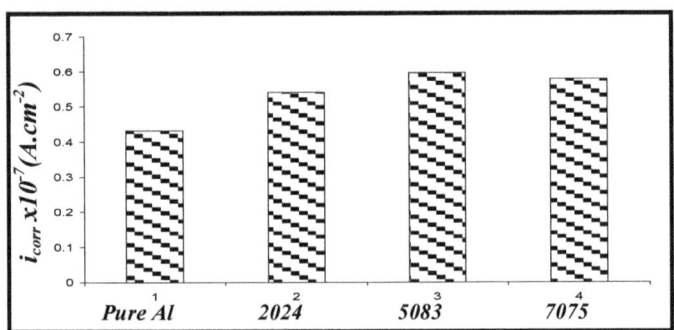

Figure (5-37) : *Values of i$_{corr}$ plotted for pure Al and its alloys in pH=9 in the presence of 0.05 mol.dm^{-3} CH$_3$COONa at 298 K.*

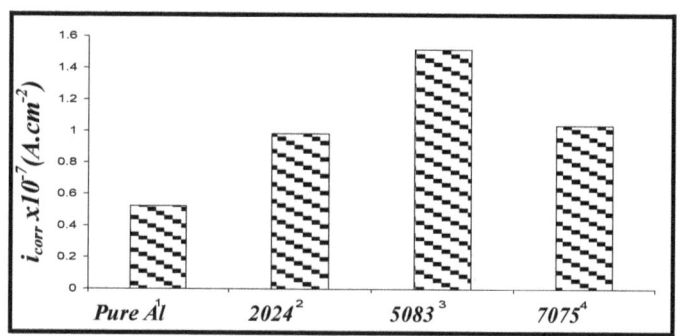

Figure (5-38) _: Values of i_corr_ plotted for pure Al and its alloys in pH=9 in the presence of 0.10 mol.dm⁻³ CH₃COONa at 298 K._

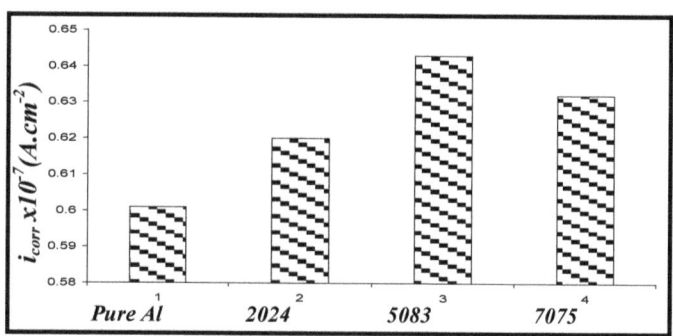

Figure (5-39) _: Values of i_corr_ plotted for pure Al and its alloys in pH=9 in the presence of 0.15 mol.dm⁻³ CH₃COONa at 298 K._

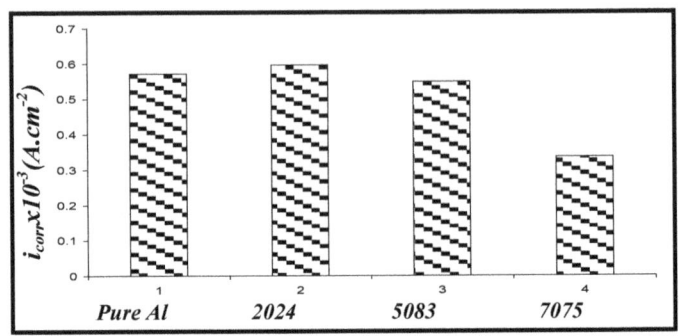

Figure (5-40) _: Values of i_{corr} plotted for pure Al and its alloys in pH=13 in the presence of $5x10^{-3}$ mol.dm^{-3} Na$_2$CrO$_4$ at 298 K._

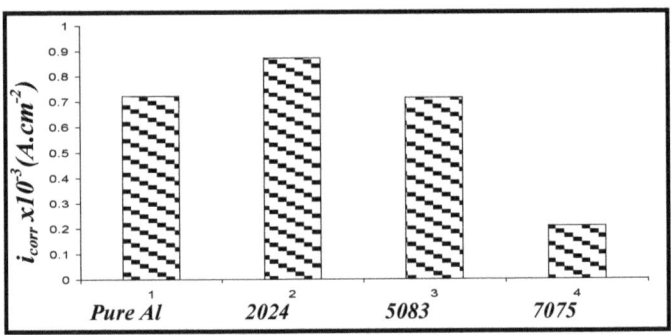

Figure (5-41) _: Values of i_{corr} plotted for pure Al and its alloys in pH=13 in the presence of $1x10^{-2}$ mol.dm^{-3} Na$_2$CrO$_4$ at 298 K._

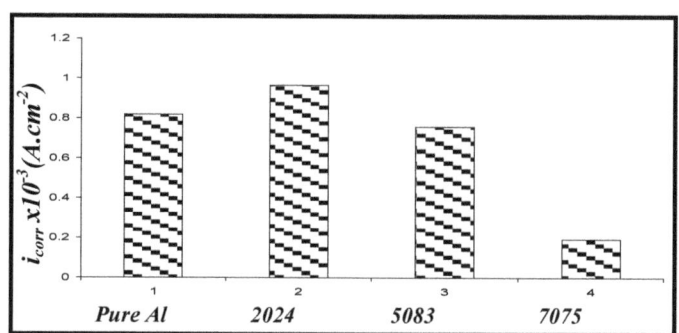

Figure (5-42) _: Values of i_{corr} plotted for pure Al and its alloys in pH=13 in the presence of $5x10^{-2}$ mol.dm^{-3} Na$_2$CrO$_4$ at 298 K._

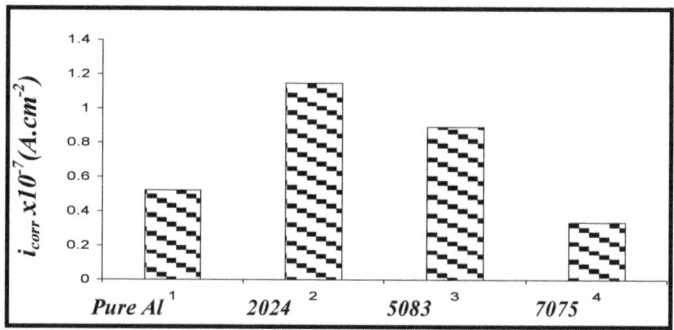

Figure (5-43) _: Values of i_{corr} plotted for pure Al and its alloys in pH=11 in the presence of $5x10^{-3}$ mol.dm^{-3} Na$_2$CrO$_4$ at 298 K._

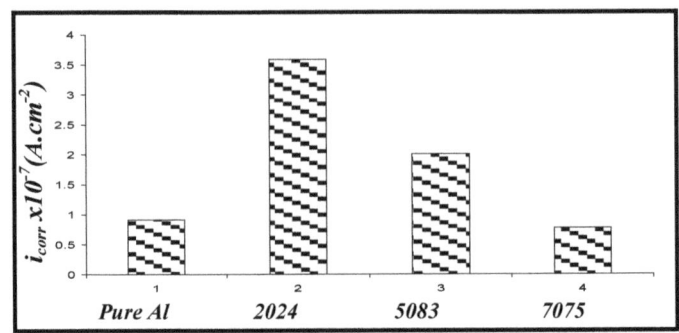

Figure (5-44) : *Values of i_{corr} plotted for pure Al and its alloys in pH=11 in the presence of $1x10^{-2}$ mol.dm^{-3} Na$_2$CrO$_4$ at 298 K.*

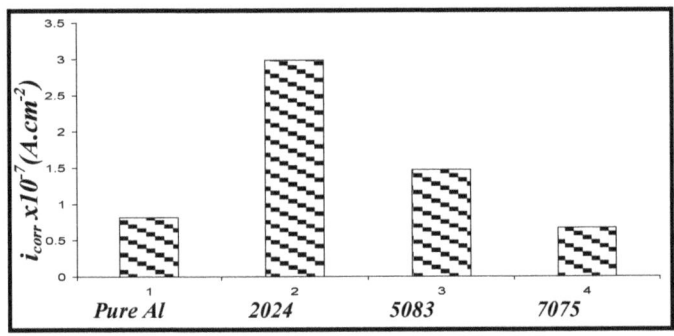

Figure (5-45) : *Values of i_{corr} plotted for pure Al and its alloys in pH=11 in the presence of $5x10^{-2}$ mol.dm^{-3} Na$_2$CrO$_4$ at 298 K.*

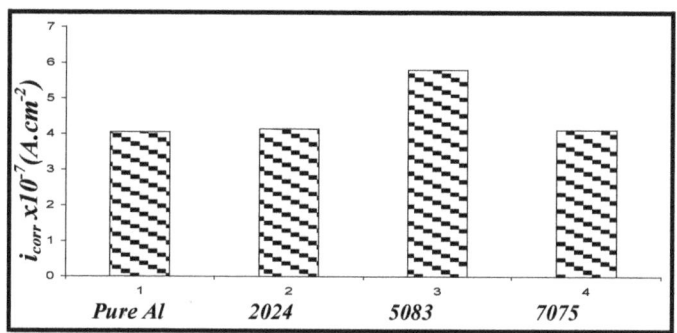

Figure (5-46) : *Values of* i_{corr} *plotted for pure Al and its alloys in pH=9 in the presence of* $5x10^{-3}$ *mol.dm*$^{-3}$ *Na_2CrO_4 at 298 K.*

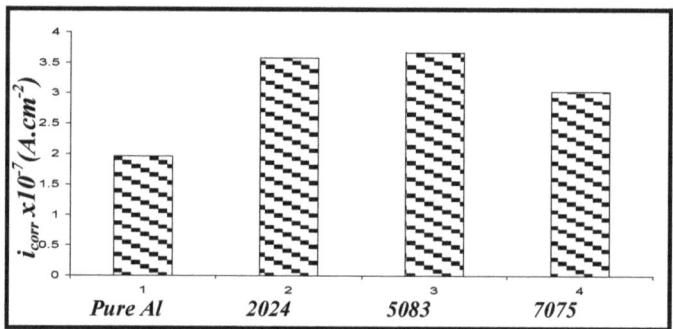

Figure (5-47) : *Values of* i_{corr} *plotted for pure Al and its alloys in pH=9 in the presence of* $1x10^{-2}$ *mol.dm*$^{-3}$ *Na_2CrO_4 at 298 K.*

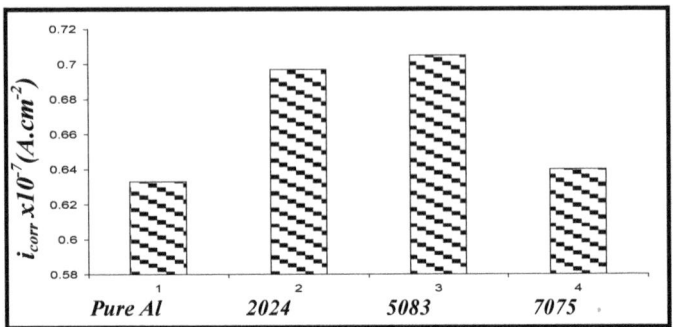

Figure (5-48) : *Values of i_{corr} plotted for pure Al and its alloys in pH=9 in the presence of $5x10^{-2}$ mol.dm^{-3} Na$_2$CrO$_4$ at 298 K.*

5-5 Tafel Slope (b) And Transfer Coefficients (α)

Tables (5-1) to (5-12) and (5-13) to (5-24) show the influence of the temperature (T) and concentration of the additives on the cathodic (b_c) and anodic (b_a) Tafel slopes which have been obtained from the polarization curves of the pure Al and its alloys in NaOH solution over the temperature range (298 – 313)K. Values of the transfer coefficients for the cathodic and anodic processes have been calculated from the corresponding cathodic and anodic Tafel slopes using the relations in (3-1) and (3-2) equations[125, 35].

A cathodic Tafel slope of **-0.120 V** (or of $α_c$ = 0.5) in general diagnostic of a discharge – chemical desorption mechanism for hydrogen evolution reaction of the cathode in which the proton discharge is the rate – determining step. If a chemical desorption is the rate – determining step, the rate will then be independent on the overpotential since no charge transfer occurs in such a step and the rate becomes directly proportional to the concentration or the coverage (θ) of the adsorbed hydrogen atoms, and may occur at coverage ranging from very small values to almost full surface layer formation[127]. The expected Tafel slope in such step would then be (-0.03

V.decade^{-1}) and (α_c = 2). When an electrochemical desorption becomes the rate – determining step for the hydrogen evolution reaction on the cathode, the expected value of b_c is (-0.05 V.decade^{-1}) and (α_c = 1.5). The results of the Tables indicate that the variation of the Tafel slopes and of the corresponding transfer coefficients could be interpreted in terms of the variation of rate – determining step from charge transfer process to either chemical – desorption or to electrochemical desorption. The variation of the anodic transfer coefficients (α_a) may be attributed to the variation of the rate – determining step in the metal dissolution reaction. A change in the mechanism as well as in the rate – determining step, cannot be ignored throughout the anodic processes[11].

5-6 Polarization Resistance

Another approach to the problem of the electrochemical corrosion rate measurement is to apply only a small potential difference to the specimen and then measure the current density. The potential – current density plot is approximately linear in the region within ± 10mV of the corrosion potential. The slope of this plot in terms of potential divided by current density has the units of resistance area and is often called the polarization resistance (R_p). The measurement of polarization resistance has very similar requirements to the measurement of full polarization curves and it is particularly useful as a method to rapidly identifying corrosion up-seting and initiating remedial action[121].

The results of Tables (5-1) to (5-12) indicate the effect of sodium acetate additive as follow :

1- In pH=13, the values of the polarization resistance in the absence of additives are generally greater than in the presence of additive of pure Al and its alloys and the greatest value of R_p is observed at 0.1 mol.dm^{-3} CH$_3$COONa and according to the sequence:

$$R_p \quad 7075 > pure\ Al > 2024 > 5083$$

2- In the pH=11 and 9, the values of the polarization resistance in the presence of CH$_3$COONa are generally greater than in the absence of the additive for pure Al and its alloys and the greatest value of R_p is observed at 0.05 mol.dm^{-3} CH$_3$COONa and according to the sequence :

$$In\ pH=11 \quad R_p \quad 2024 > pure\ Al > 7075 > 5083$$
$$In\ pH=9 \quad R_p \quad pure\ Al > 7075 > 5083 > 2024$$

While the Tables (5-13) to (5-24) indicate the effect of the presence of sodium chromate as follow :

1- In pH=13, the values of R_p in the absence of Na$_2$CrO$_4$ are generally greater than in the presence of the additive for pure Al and its alloys and the greatest value of R_p is observed at 5x10^{-3} mol.dm^{-3} Na$_2$CrO$_4$ and according to the sequence :

$$R_p \quad 7075 > pure\ Al > 5083 > 2024$$

2- In pH=11, the values of R_p in the presence of Na$_2$CrO$_4$ are generally greater than in the absence of the additive for pure Al and its alloys and the greatest value is observed at 5x10^{-3} mol.dm^{-3} Na$_2$CrO$_4$ and according to the sequence :

$$R_p \quad 7075 > pure\ Al > 5083 > 2024$$

3- In pH=9, the values of R_p in the presence of Na$_2$CrO$_4$ are generally greater than in the absence of the additive and the greatest value of R_p is observed at 5x10^{-2} mol.dm^{-3} Na$_2$CrO$_4$ and according to the sequence :

$$R_p \quad pure\ Al > 7075 > 5083 > 2024$$

176

5-7 Protection Efficiency (P%)

The corrosion current densities in the presence and absence of the additives in the corrosion medium have been used to determine the protection efficiency (P%) using the relation :

$$P\% = 100\left[1 - \frac{i_2}{i_1}\right] \quad \ldots\ldots(5\text{-}1)$$

where i_1 and i_2 are the corrosion current densities in the absence and presence of additive in the corrosion medium respectively.

A positive value of P% indicates the inhibition of corrosion by the added additives while a negative value of P% implies corrosion stimulation or corrosion acceleration.

The results of Tables (5-25) to (5-32) are summarize in the following:

1- Values of (P%) are negative for pure Al and its alloys in 0.1 mol.dm^{-3} NaOH solution (pH=13) in the presence of CH$_3$COONa and Na$_2$CrO$_4$ with three experimental concentration.

The decrease in inhibitive efficiency with the concentration increasing of the alkali may be attributed to the high rate of evolution of hydrogen in concentrated alkalies which may interfere with the absorption of the inhibitor on the metal surface.

2- In pH=11, the (P%) varies for pure Al and its alloys in the presence of sodium acetate at three experimental concentration.

Fig. (5-49) show the effect of the concentration of sodium acetate on (P%) for pure Al and its alloys at 298K. the order of the variation of P% values at pH=11 in the presence of sodium acetate in the sequence :

$$P\% \qquad 2024 > pure\ Al > 7075 > 5083$$

While in the presence of sodium chromate as an inhibitor in pH=11, gives better inhibition for pure Al and its alloys as shown in Fig. (5-51) with three experimental concentration of sodium chromate at 298K.

3- The presence of sodium acetate in pH=9 gives good protection efficiency (P%) for pure Al and its alloys and higher than the corresponding values in pH=11. Fig. (5-50) show the effect of sodium acetate concentration on the (P%) in PH=9 at 298K.

4- The presence of sodium chromate in pH=9 gives good protection efficiency for pure Al and its alloys but lower than that is observed in the pH=11. Fig. (5-52) shows the effect of the concentration of sodium chromate on the P% at constant PH and temperature.

5- The effect of temperature on inhibitor efficiency is shown in Figs. (5-53) to (5-56). It is observed that the extent of inhibition , in general, decrease with the rise in temperature.

Table (5-25) : *Protection efficiencies (P%) for the corrosion of pure Al in NaOH solution in presence of (0.05, 0.10, 0.15 mol.dm^{-3}) sodium acetate at four temperatures.*

The medium (pH)	T (K)	P%		
		0.05 mol.dm^{-3}	0.10 mol.dm^{-3}	0.15 mol.dm^{-3}
13	298	-78448.38	-55383.87	-73125.80
	303	-52324.24	-44647.47	-5464.47
	308	-34086.04	-31353.48	-36993.02
	313	-24160.86	-18450.72	-22740.57
11	298	48.25	27.73	-58.17
	303	60.03	20.53	-7.540
	308	54.08	-8.110	-21.28
	313	37.96	-20.50	-71.26
9	298	94.30	89.63	93.47
	303	94.09	88.85	93.11
	308	93.62	87.49	92.36
	313	92.39	86.70	91.40

Table (5-26) *: Protection efficiencies (P%) for the corrosion of 2024 alloy in NaOH solution in presence of (0.05, 0.10, 0.15 mol.dm⁻³) sodium acetate at four temperatures.*

The medium (pH)	T (K)	P%		
		0.05 mol.dm⁻³	0.10 mol.dm⁻³	0.15 mol.dm⁻³
13	298	-10976.92	-6207.69	-11233.33
	303	-10240.42	-6453.19	-11282.97
	308	-10427.27	-8409.09	-9990.90
	313	-15069.49	-10730.50	-10357.62
11	298	98.57	97.86	97.56
	303	98.42	97.82	97.50
	308	98.52	97.63	97.47
	313	98.55	97.76	96.84
9	298	99.95	99.95	99.94
	303	99.95	99.92	99.94
	308	99.96	99.92	99.94
	313	99.96	99.92	99.94

Table (5-27) *: Protection efficiencies (P%) for the corrosion of 5083 alloy in NaOH solution in presence of (0.05, 0.10, 0.15 mol.dm⁻³) sodium acetate at four temperatures.*

The medium (pH)	T (K)	P%		
		0.05 mol.dm⁻³	0.10 mol.dm⁻³	0.15 mol.dm⁻³
13	298	-84066.66	-99400.0	-78900
	303	-59791.30	-79791.30	-61965.21
	308	59008.91	-86335.64	-68117.82
	313	-59813.04	-79813.04	-91813.04
11	298	-136.613	-162.92	-189.24
	303	-128.36	-185.50	-280.74
	308	-100.00	-220.10	-300.08
	313	-73.310	-200.00	-300.07
9	298	93.90	84.52	93.43
	303	93.31	83.93	92.86
	308	93.25	83.12	92.35
	313	92.34	80.30	91.38

Table (5-28) : _Protection efficiencies (P%) for the corrosion of 7075 alloy in NaOH solution in presence of (0.05, 0.10, 0.15 mol.dm^{-3}) sodium acetate at four temperatures._

The medium (pH)	T (K)	P%		
		0.05 mol.dm^{-3}	0.10 mol.dm^{-3}	0.15 mol.dm^{-3}
13	298	-179668.78	-85449.13	-253657.20
	303	-123471.42	-61685.71	-186685.71
	308	-60044.92	-64972.46	-77146.37
	313	-49788.88	-69011.11	-63344.44
11	298	-66.61	-149.91	-288.76
	303	-76.06	-242.64	-280.74
	308	-79.53	-260.00	-309.36
	313	-66.66	-266.66	-333.33
9	298	94.43	90.01	93.92
	303	93.96	89.84	93.76
	308	93.26	87.42	92.95
	313	92.98	85.70	92.10

Table (5-29) : _Protection efficiencies (P%) for the corrosion of pure Al in NaOH solution in presence of (5x10^{-3}, 1x10^{-2}, 5x10^{-2} mol.dm^{-3}) sodium chromate at four temperatures._

The medium (pH)	T (K)	P%		
		5x10^{-3} mol.dm^{-3}	1x10^{-2} mol.dm^{-3}	5x10^{-2} mol.dm^{-3}
13	298	-14566.66	-18412.82	-20874.35
	303	-13304.25	-19836.17	-20751.06
	308	-13663.63	-18172.72	-19081.81
	313	-14323.72	-18732.20	-19832.20
11	298	99.99	99.98	99.98
	303	99.98	99.98	99.98
	308	99.99	99.97	99.98
	313	99.99	99.97	99.98
9	298	99.62	99.81	99.94
	303	99.59	99.79	99.94
	308	99.66	99.77	99.95
	313	99.66	99.73	99.94

Table (5-30) _: Protection efficiencies (P%) for the corrosion of 2024 alloy in NaOH solution in presence of (5x10^{-3}, 1x10^{-2}, 5x10^{-2} mol.dm^{-3}) sodium chromate at four temperatures._

The medium (pH)	T (K)	P%		
		5x10^{-3} mol.dm^{-3}	1x10^{-2} mol.dm^{-3}	5x10^{-2} mol.dm^{-3}
13	298	-96190.32	-140706.45	-155545.16
	303	-69495.95	-95758.58	-102021.21
	308	-45888.37	-58911.62	-66702.32
	313	-25784.05	-32392.75	-34537.68
11	298	98.70	95.95	98.63
	303	99.22	97.81	98.13
	308	99.24	97.73	98.10
	313	99.13	97.66	98.13
9	298	56.45	62.33	92.66
	303	49.43	58.63	92.65
	308	44.65	57.43	92.32
	313	33.12	55.84	90.22

Table (5-31) _: Protection efficiencies (P%) for the corrosion of 5083 alloy in NaOH solution in presence of (5x10^{-3}, 1x10^{-2}, 5x10^{-2} mol.dm^{-3}) sodium chromate at four temperatures._

The medium (pH)	T (K)	P%		
		5x10^{-3} mol.dm^{-3}	1x10^{-2} mol.dm^{-3}	5x10^{-2} mol.dm^{-3}
13	298	-91566.66	-118900.0	-126233.33
	303	-67834.78	-82291.30	-87182.60
	308	-66137.62	-79404.95	-88216.83
	313	-61986.95	-73639.13	-85291.30
11	298	98.97	97.70	98.31
	303	99.03	97.68	98.38
	308	99.14	97.66	98.44
	313	99.22	97.50	98.60
9	298	40.78	62.54	92.80
	303	39.32	58.82	92.64
	308	30.37	54.95	92.22
	313	24.26	48.18	90.15

Table (5-32) : *Protection efficiencies (P%) for the corrosion of 7075 alloy in NaOH solution in presence of ($5x10^{-3}$, $1x10^{-2}$, $5x10^{-2}$ mol.dm^{-3}) sodium chromate at four temperatures.*

The medium (pH)	T (K)	P% $5x10^{-3}$ mol.dm^{-3}	P% $1x10^{-2}$ mol.dm^{-3}	P% $5x10^{-2}$ mol.dm^{-3}
13	298	-194119.65	-21287.28	-112038.72
	303	-142828.57	-218900.0	-92042.85
	308	-68015.94	-53378.26	-41204.34
	313	-55900.00	-48344.44	-37233.33
11	298	99.46	98.75	98.92
	303	99.40	98.84	98.98
	308	99.45	98.83	99.06
	313	99.48	98.86	99.19
9	298	60.45	70.91	93.84
	303	58.83	69.72	93.63
	308	57.27	68.66	93.06
	313	52.77	65.84	91.94

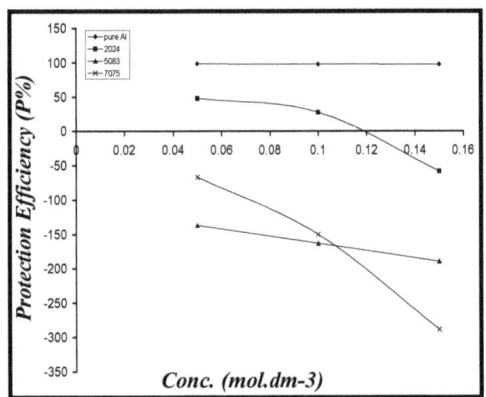

Fig. (5-49) : *Effect of inhibitor concentration (CH$_3$COONa) on the protection efficiency for pure Al and its alloys in pH=11 at 298K.*

182

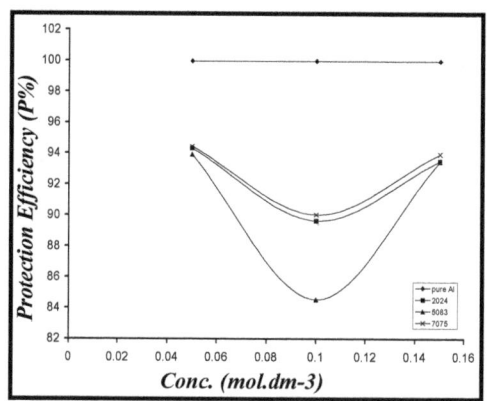

Fig. (5-50) : *Effect of inhibitor concentration
(CH₃COONa) on the protection efficiency for
pure Al and its alloys in pH=9 at 298K.*

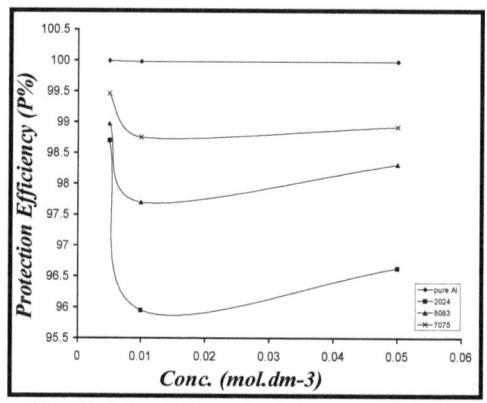

Fig. (5-51) : *Effect of inhibitor concentration
(Na₂CrO₄) on the protection efficiency for
pure Al and its alloys in pH=11 at 298K.*

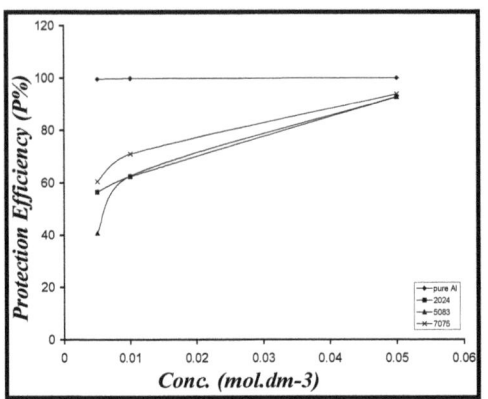

Fig. (5-52) *: Effect of inhibitor concentration*
(Na₂CrO₄) on the protection efficiency for
pure Al and its alloys in pH=9 at 298K.

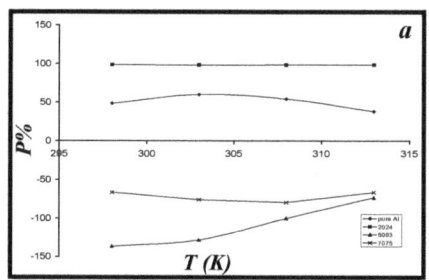

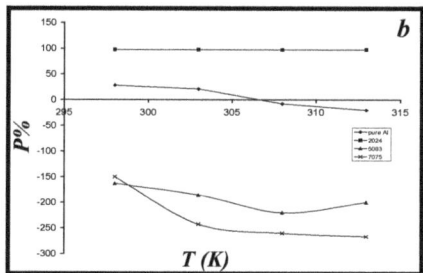

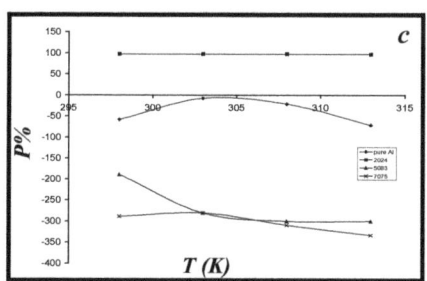

Fig. (5-53) _: The variation of the protection efficiencies_

(P%) with temperature (T) for pure Al and its

alloys in pH=11 in presence of

(a)0.05 mol.dm^{-3}CH$_3$COONa,

(b)0.10 mol.dm^{-3} CH$_3$COONa,

(c)0.15 mol.dm^{-3} CH$_3$COONa.

185

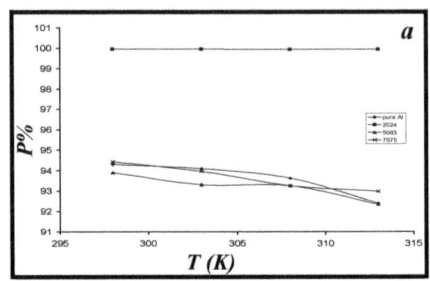

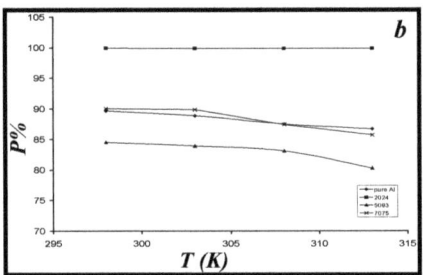

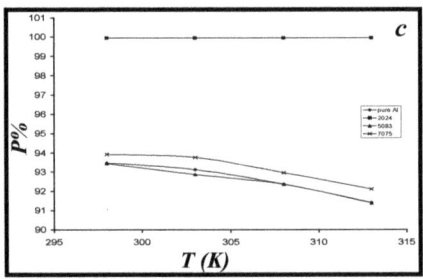

Fig. (5-54) _: The variation of the protection efficiencies_
(P%) with temperature (T) for pure Al and its
alloys in pH=9 in presence of
(a)0.05 mol.dm⁻³ CH₃COONa,
(b)0.10 mol.dm⁻³ CH₃COONa,
(c)0.15 mol.dm⁻³ CH₃COONa.

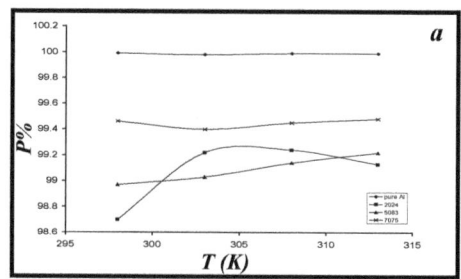

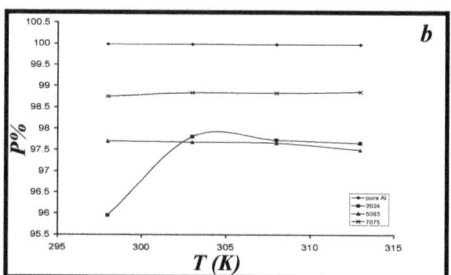

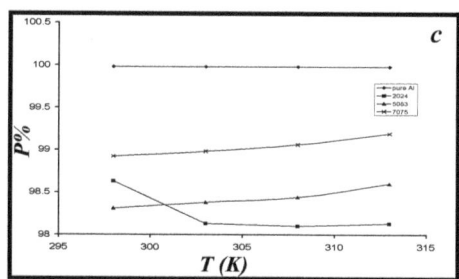

Fig. (5-55) *: The variation of the protection efficiencies*
(P%) with temperature (T) for pure Al and its
alloys in pH=11 in presence of
(a)$5x10^{-3}$ mol.dm^{-3}Na$_2$CrO$_4$,
(b)$1x10^{-2}$ mol.dm^{-3}Na$_2$CrO$_4$,
(c)$5x10^{-2}$ mol.dm^{-3}Na$_2$CrO$_4$.

187

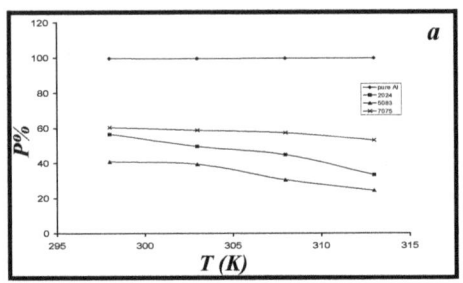

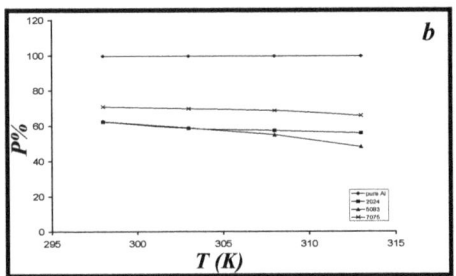

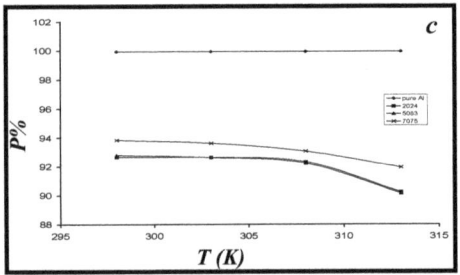

Fig. (5-56) *: The variation of the protection efficiencies*
(P%) with temperature (T) for pure Al and its
alloys in pH=9 in presence of
(a) $5x10^{-3}$ mol.dm^{-3}Na$_2$CrO$_4$,
(b) $1x10^{-2}$ mol.dm^{-3}Na$_2$CrO$_4$,
(c) $5x10^{-2}$ mol.dm^{-3}Na$_{)2}$CrO$_4$.

188

5-8 General Discussion Of Inhibition Action

According to Straumanis and Brakss[161] the initial reactions in aqueous solutions of sodium hydroxide may be represented as:

$$Al + 3H_2O \rightarrow Al(OH)_3 + 3/2H_2 \qquad .... (i)$$

This is then followed by the reaction

$$Al(OH)_3 + OH^- \rightarrow AlO_2^- + 2H_2O \quad ..... (ii)$$

Thus, the net reaction is

$$Al + NaOH + H_2O \rightarrow NaAlO_2 + 3/2H_2 \quad ...... (iii)$$

According to Putilova et al [162], reaction (ii) may also be represented as:

$$Al(OH)_3 + NaOH + 2H_2O \rightarrow Na^+ + [Al(OH)_4 2H_2O] \quad ..... (iv)$$

As aluminium hydroxide gets precipitated at pH values higher than (4.1) and lower than (10), there is no possibility of deceleration of attack due to an oxide film formation beyond these values. That is why the corrosion of aluminium proceeds unabated in alkaline solution.

According to Antropov[163], for a corrosion process occurring at a very negative potential, e.g. dissolution of aluminium in alkaline media, all organic compounds will be ousted from the electrical double layer and effective inhibition will be difficult.

This work uses sodium acetate as an organic inhibitor. The inhibiting action achieved by organic compounds is usually attributed to interactions by adsorption between the inhibitor and the metal surface.

Adsorption can be of a purely physical nature by means of electrostatic or Vande Waals forces, which are easily removed from the surface, or a chemical nature which forms chemical compounds.

They undergo a chemisorptions process which involves a charge sharing or a charge transfer process. This takes place more slowly than the physical adsorption, but the bonding achieved is stronger, making desorption considerably more difficult.

Inhibitor efficiency is higher for a compound which can donate electrons easily for the molecular site of adsorption and corresponds to high electron density at the presumed adsorption center in the molecular.

Most organic inhibitors are compounds with at least one polar function, the polar function is regarded as the reaction center for the establishment of the chemisorbed bond, whose strength is determined by the electron density of the atom acting as the reaction center[164].

$$: O :$$
$$\|$$
$$H_3C\!\!-\!\!C\!\!-\!\!O\!\!:\!-$$

(acetate ion)

Generally, organic compounds are adsorbed on the metal surface and interface with either cathodic or anodic reaction occurring at the adsorption site. The bulky molecules limit the diffusion of oxygen to the surface or they trap the metal ions on the surface, reducing the rate of dissolution, while inorganic oxidizing substance that promote the passivity on the surface by shifting the corrosion potential in the noble direction.

Acetate is classified in the class (I), while Pourbax et al. have quoted work by Charlot which would place acetate within class (II) (section*5-1-2*).

Acetate ions gave rise to small irregularly shaped pits with clear crystallographic features. This observation suggests that the acetate ion is neither a passivator nor a blocking inhibitor by forming insoluble precipitates. Its action must be either to isolate the surface from other ions by preferential adsorption or to act as a buffer[72].

Chromates are efficient inhibitors of the corrosion of aluminium and its alloys in near – neutral aqueous environments containing aggressive anions.

Additionally; chromates in the form of inhibitive pigments are incorporated into primer paints for the protection of aluminium alloys particularly at damaged regions. Concerning the mechanism of inhibition of corrosion of aluminium, several studies show general agreement in the importance of Cr^{6+} ion reduction to hydrated Cr_2O_3, but it differs in the proposed location of the solid material.

Edeleanu and Evans[165] suggest that a redox reaction between Cr^{6+} ions and aluminium metal, however revealed, occurs, forming aluminium and solid chromic oxide:

$$2Al\ +\ 3H_2O\ \rightarrow\ Al_2O_3\ +\ 6H^+\ +\ 6e$$
$$2CrO_4^{2-}\ +\ 10H^+\ +\ 6e\ \rightarrow\ Cr_2O_3\ +\ 5H_2O$$

The development of solid Cr_2O_3 generally over the surface was assumed by Pryor et. al.[166] and substantiated by later X – ray photo – electron studies (XPS) of aluminium specimens, supporting air – formed film, which had been immersed in chromate electrolyte.

However, while XPS can provide information on valency states in the near surface layer, spatial resolution, of importance in locating the exact sites of electrochemical processes, is poor. Abd Rabbo et al.[150] utilized secondary ion mass spectrometry (SIMS), offering high secondary ion sensitivity and good spatial resolution, to study aluminium supporting relatively thick barrier – type anodic films, which had been immersed in chromate electrolytes for various times.

An initial, general yield of instrumentally induced Cr^+ secondary ions were observed, which were rapidly decreased with sputtering to give a localized distribution of Cr^+ ions through the film to the aluminium / film interface; the latter are associated with pitting sites in aluminium substrates. Although the valency states of particular chromium – containing species could not be determined, the data were taken to indicate the initial general presence of chromate species which had penetrated the anodic alumina film and the local development, at likely flaw sites, of solid Cr_2O_3 plugs. The local development of $Al(OH)_3$ plugs, at anodic sites, the precipitation of which is catalyzed by chromate species, was not discounted.

The data of various workers, using different experimental conditions and examination techniques, suggest that Cr(III) is present predominantly on specimens supporting air – formed films (after appropriate immersion times) whilst Cr(VI) is detected in the outer regions of specimens supporting relatively thick barrier – type films, with a local presence of Cr(III).

In the limit of this work, it was observed that the best inhibition by using sodium acetate as an organic inhibitor getting at pH=9 with $5x10^{-2}$ mol.dm^{-3} of this inhibitor, and according to the sequence:

2024 > pure Al > 7075 > 5083

While the best inhibition is by using sodium chromate as an inorganic inhibitor getting at pH=11 with $5x10^{-3}$ mol.dm^{-3} of this inhibitor for pure Al and its alloys.

Chapter Six: Thermodynamic and Kinetic Fuctions

6-1 Thermodynamic Of Corrosion

Thermodynamics, the science of energy changes, has been widely applied to corrosion studies for many years. The change in free energy (ΔG) is a direct measure of the work capacity or maximum electric energy available from a system. Chemical and corrosion reactions behave in exactly the same fashion. Thermodynamic laws tell us that there is a strong tendency for high energy state in a system to transform into low energy state. It is this tendency of metals to recombine with components of the environment that leads to the phenomenon known as corrosion[167].

The free – energy change accompanying an electrochemical reaction can be calculated by the following equation[35, 126] :

$$\Delta G = -nFE \qquad \qquad ......(6-1)$$

where ΔG is the free – energy change, n is the number of electrons involved in the reaction, F is the Faraday constant, and E equals the cell potential ($E = E_{corr}$).

From the value of ΔG at several temperatures, the change in the entropy (ΔS) of corrosion process could be derived according to the well – known thermodynamic relation :

$$\Delta S = -\frac{d(\Delta G)}{dT} \qquad \qquad .....(6-2)$$

Values of ΔG are usually plotted against temperature (T); thus at any temperature the value of $-d(\Delta G)/dT = \Delta S$ which corresponds to the slope of the $(-\Delta G)$ versus (T) plot at that temperature, as shown in Figs. (6-3) to (6-5).

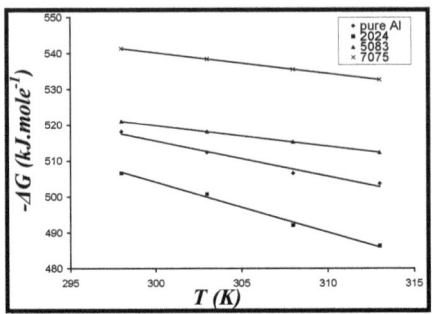

Fig. (6-3) : *The variation of the (-ΔG)with temperature (T) for corrosion of pure Al and its alloys in pH=13 without additives.*

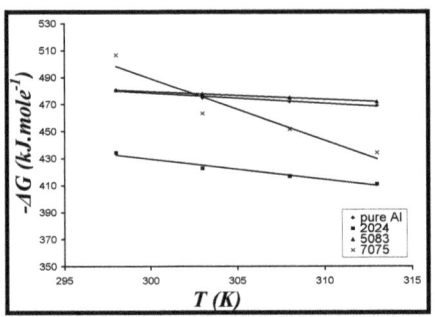

Fig. (6-4) : *The variation of the (-ΔG)with temperature (T) for corrosion of pure Al and its alloys in pH=11 without additives.*

194

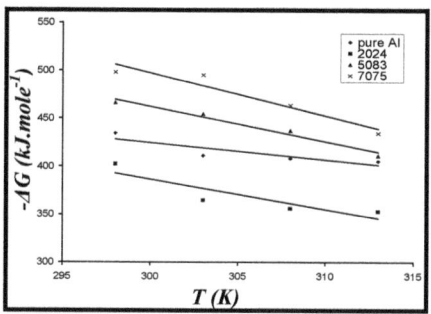

Fig. (6-5) : _The variation of the (-ΔG)with_
temperature (T) for corrosion of pure Al
and its alloys in pH=9 without additives.

The change in free energy, ΔG, is related to ΔH, the change in the
enthalpy, and ΔS, the change in entropy of the corrosion reaction at a constant
temperature, T, by the equation[168, 169] :

$$\Delta G = \Delta H - T\Delta S \qquad \ldots\ldots(6-3)$$

Tables (6-1) to (6-10) give values of the thermodynamic quantities for
the corrosion of aluminium and its alloys in NaOH solutions with three values
of pH (13, 11, and 9) in the absence and presence of additives (NaCl,
CH_3COONa, and Na_2CrO_4) at four temperatures.

Tables (6-1) : *The thermodynamic quantities for pure Al and its alloy in NaOH solution at three value of pH without additives at four temperatures.*

Elect.	T (K)	$-\Delta G$ (kJ.mole^{-1})			ΔS (J.mol^{-1}.K^{-1})			ΔH (kJ.mole^{-1})		
		13	11	9	13	11	9	13	11	9
Pure Al	298	518.17	480.54	434.22	984.2	752.6	1794.7	-224.9	-256.4	100.3
	303	512.38	474.75	411.06	-	-	-	-	-	-
	308	506.59	471.85	408.16	-	-	-	-	-	-
	313	503.69	468.96	405.27	-	-	-	-	-	-
2024	298	506.59	434.22	402.38	1389.5	1505.3	3126.4	-92.66	14.27	529.1
	303	500.80	422.64	364.74	-	-	-	-	-	-
	308	492.11	416.85	356.06	-	-	-	-	-	-
	313	486.32	411.06	353.16	-	-	-	-	-	-
5083	298	521.06	480.54	466.06	578.9	578.9	3647.4	-348.8	-308.2	620.7
	303	518.17	477.64	454.48	-	-	-	-	-	-
	308	515.27	474.75	437.11	-	-	-	-	-	-
	313	512.38	471.85	411.06	-	-	-	-	-	-
7075	298	541.33	506.59	497.90	578.9	4573.8	4458	-369.0	857.6	866.3
	303	538.43	463.17	495.01	-	-	-	-	-	-
	308	535.54	451.59	463.17	-	-	-	-	-	-
	313	532.64	434.22	434.22	-	-	-	-	-	-

Tables (6-2) : *The thermodynamic quantities for pure Al and its alloy in pH=13 in presence of NaCl at four temperatures.*

Elect.	T (K)	$-\Delta G$ (kJ.mole^{-1})			ΔS (J.mol^{-1}.K^{-1})			ΔH (kJ.mole^{-1})		
		10^{-3}	10^{-2}	0.1	10^{-3}	10^{-2}	0.1	10^{-3}	10^{-2}	0.1
Pure Al	298	648.43	645.54	633.96	810.5	1215.8	578.9	-407.0	-283.4	-461.7
	303	645.54	639.75	631.07	-	-	-	-	-	-
	308	639.75	631.07	628.17	-	-	-	-	-	-
	313	636.86	628.17	625.28	-	-	-	-	-	-
2024	298	552.91	518.17	515.27	5442.2	3300	3068.5	1069	465.2	398.9
	303	518.17	503.69	482.54	-	-	-	-	-	-
	308	489.22	486.32	474.75	-	-	-	-	-	-
	313	471.85	468.96	466.06	-	-	-	-	-	-
5083	298	599.22	596.33	578.96	984.2	1273.7	1389.5	-305.9	-216.9	-165.0
	303	593.43	590.54	573.17	-	-	-	-	-	-
	308	587.64	587.64	564.48	-	-	-	-	-	-
	313	584.75	576.06	558.70	-	-	-	-	-	-
7075	298	602.12	599.22	596.33	1042.1	926.3	926.3	-291.6	-323.2	-320.3
	303	599.22	596.33	593.43	-	-	-	-	-	-
	308	590.54	593.43	590.54	-	-	-	-	-	-
	313	587.64	584.75	581.85	-	-	-	-	-	-

Tables (6-3) : *The thermodynamic quantities for pure Al and its alloy in pH=11 in presence of NaCl at four temperatures.*

Elect.	T (K)	-ΔG (kJ.mole⁻¹)			ΔS (kJ.mol⁻¹.K⁻¹)			ΔH (kJ.mole⁻¹)		
		10^{-3}	10^{-2}	0.1	10^{-3}	10^{-2}	0.1	10^{-3}	10^{-2}	0.1
Pure Al	298	613.70	628.17	616.59	1679	578.9	1505.3	-113.6	-455.9	-168.1
	303	607.91	625.28	605.01	-	-	-	-	-	-
	308	602.12	622.38	599.22	-	-	-	-	-	-
	313	587.64	619.49	593.43	-	-	-	-	-	-
2024	298	413.95	495.01	434.22	1563.2	4342.2	2952.7	51.82	798.9	445.4
	303	405.27	445.80	413.95	-	-	-	-	-	-
	308	396.59	437.11	405.27	-	-	-	-	-	-
	313	390.80	425.53	387.90	-	-	-	-	-	-
5083	298	529.75	564.48	550.01	1273.7	752.6	926.3	-150.3	-340.3	-274.1
	303	518.17	561.59	541.33	-	-	-	-	-	-
	308	515.27	558.70	538.43	-	-	-	-	-	-
	313	509.48	552.91	535.54	-	-	-	-	-	-
7075	298	547.12	587.64	552.91	1447.4	2084.2	752.6	-115.9	33.39	-328.8
	303	544.22	573.17	547.12	-	-	-	-	-	-
	308	532.64	564.48	544.22	-	-	-	-	-	-
	313	526.85	555.80	541.33	-	-	-	-	-	-

Tables (6-4) : *The thermodynamic quantities for pure Al and its alloy in pH=9 in presence of NaCl at four temperatures.*

Elect.	T (K)	-ΔG (kJ.mole⁻¹)			ΔS (J.mol⁻¹.K⁻¹)			ΔH (kJ.mole⁻¹)		
		10^{-3}	10^{-2}	0.1	10^{-3}	10^{-2}	0.1	10^{-3}	10^{-2}	0.1
Pure Al	298	648.43	228.69	205.53	5673.8	1852.6	1100	1042.1	323.2	122.3
	303	633.96	211.32	208.42	-	-	-	-	-	-
	308	619.49	205.53	211.32	-	-	-	-	-	-
	313	558.70	199.74	222.90	-	-	-	-	-	-
2024	298	405.27	-2.894	40.52	3647.6	1215.8	578.9	681.5	364.9	131.7
	303	390.80	-8.684	43.422	-	-	-	-	-	-
	308	364.74	-17.36	46.317	-	-	-	-	-	-
	313	353.16	-20.26	49.211	-	-	-	-	-	-
5083	298	521.06	205.53	228.69	5558	1679	1736.9	1135.2	294.5	288.6
	303	495.01	199.74	231.58	-	-	-	-	-	-
	308	477.64	185.26	240.27	-	-	-	-	-	-
	313	434.22	182.37	254.74	-	-	-	-	-	-
7075	298	584.75	622.38	306.85	3415.8	8452.8	3300	432.9	1896.3	676.5
	303	578.96	616.59	309.74	-	-	-	-	-	-
	308	555.80	532.64	344.48	-	-	-	-	-	-
	313	535.54	509.48	350.27	-	-	-	-	-	-

Tables (6-5) : *The thermodynamic quantities for pure Al and its alloy in pH=13 in presence of CH_3COONa at four temperatures.*

Elect.	T (K)	$-\Delta G$ (kJ.mole^{-1})			ΔS (kJ.mol^{-1}.K^{-1})			ΔH (kJ.mole^{-1})		
		0.05	0.10	0.15	0.05	0.10	0.15	0.05	0.10	0.15
Pure Al	298	636.86	633.96	648.43	578.9	578.9	1157.9	-464.6	-461.7	-303.6
	303	633.96	631.07	645.54	-	-	-	-	-	-
	308	631.07	628.17	639.75	-	-	-	-	-	-
	313	628.17	625.28	631.07	-	-	-	-	-	-
2024	298	526.85	523.96	521.06	3357.9	1910.5	2894.8	473.5	45.22	341.3
	303	509.48	518.17	509.48	-	-	-	-	-	-
	308	489.22	509.48	495.01	-	-	-	-	-	-
	313	477.64	495.01	477.64	-	-	-	-	-	-
5083	298	596.33	596.33	587.64	578.9	578.9	810.5	-424.1	-424.1	-346.2
	303	593.43	593.43	590.54	-	-	-	-	-	-
	308	590.54	590.54	596.33	-	-	-	-	-	-
	313	587.64	587.64	599.22	-	-	-	-	-	-
7075	298	605.01	613.70	616.59	984.2	1042.1	984.2	-311.7	-303.1	-323.3
	303	602.12	610.80	610.80	-	-	-	-	-	-
	308	596.33	602.12	605.01	-	-	-	-	-	-
	313	590.54	599.22	602.12	-	-	-	-	-	-

Tables (6-6) : *The thermodynamic quantities for pure Al and its alloy in pH=11 in presence of CH_3COONa at four temperatures.*

Elect.	T (K)	$-\Delta G$ (kJ.mole^{-1})			ΔS (J.mol^{-1}.K^{-1})			ΔH (kJ.mole^{-1})		
		0.05	0.10	0.15	0.05	0.10	0.15	0.05	0.10	0.15
Pure Al	298	599.22	631.07	622.38	926.3	2489.5	578.9	-323.2	110.6	-450.1
	303	596.33	625.28	619.49	-	-	-	-	-	-
	308	593.43	613.70	616.59	-	-	-	-	-	-
	313	584.75	593.43	613.70	-	-	-	-	-	-
2024	298	460.27	437.11	422.64	2431.6	1505.3	1794.7	264.1	11.38	111.9
	303	451.59	434.22	419.74	-	-	-	-	-	-
	308	434.22	428.43	408.16	-	-	-	-	-	-
	313	425.53	413.95	396.59	-	-	-	-	-	-
5083	298	538.43	529.75	541.33	1157.9	578.9	578.9	-193.6	-357.5	-369.1
	303	532.64	526.85	538.43	-	-	-	-	-	-
	308	526.85	523.96	535.54	-	-	-	-	-	-
	313	521.06	521.06	532.64	-	-	-	-	-	-
7075	298	558.70	550.01	544.22	1447.4	752.6	578.9	-127.4	-325.9	-371.9
	303	552.91	544.22	541.33	-	-	-	-	-	-
	308	541.33	541.33	538.43	-	-	-	-	-	-
	313	538.43	538.43	535.54	-	-	-	-	-	-

Tables (6-7) : *The thermodynamic quantities for pure Al and its alloy in pH=9 in presence of CH_3COONa at four temperatures.*

Elect.	T (K)	-ΔG (kJ.mole⁻¹)			ΔS (kJ.mol⁻¹.K⁻¹)			ΔH (kJ.mole⁻¹)		
		0.05	0.10	0.15	0.05	0.10	0.15	0.05	0.10	0.15
Pure Al	298	500.80	396.59	526.85	984.2	1852.6	1679	-207.5	155.3	-26.8
	303	495.01	379.22	512.38	-	-	-	-	-	-
	308	489.22	373.43	506.59	-	-	-	-	-	-
	313	486.32	367.64	500.80	-	-	-	-	-	-
2024	298	341.58	335.79	347.37	2257.9	984.2	2142.1	331.0	-42.55	290.9
	303	324.21	330.00	335.79	-	-	-	-	-	-
	308	315.53	324.21	324.21	-	-	-	-	-	-
	313	306.85	321.32	315.53	-	-	-	-	-	-
5083	298	460.27	364.74	448.69	1505.3	578.9	3589.5	-11.78	-192.4	620.8
	303	448.69	361.85	431.32	-	-	-	-	-	-
	308	442.90	358.95	425.53	-	-	-	-	-	-
	313	437.11	356.06	390.80	-	-	-	-	-	-
7075	298	463.17	393.69	518.17	1505.3	1852.6	2721.1	-14.68	158.2	292.6
	303	451.59	376.32	503.69	-	-	-	-	-	-
	308	445.80	370.53	497.90	-	-	-	-	-	-
	313	440.01	364.74	474.75	-	-	-	-	-	-

Tables (6-8) : *The thermodynamic quantities for pure Al and its alloy in pH=13 in presence of Na_2CrO_4 at four temperatures.*

Elect.	T (K)	-ΔG (kJ.mole⁻¹)			ΔS (J.mol⁻¹.K⁻¹)			ΔH (kJ.mole⁻¹)		
		0.005	0.01	0.05	0.005	0.01	0.05	0.005	0.01	0.05
Pure Al	298	607.91	500.80	489.22	578.9	578.9	578.9	-435.6	-328.5	-316.9
	303	605.01	497.90	486.32	-	-	-	-	-	-
	308	602.12	495.01	483.43	-	-	-	-	-	-
	313	599.22	492.11	480.54	-	-	-	-	-	-
2024	298	523.96	428.43	416.85	578.9	926.3	926.3	-351.7	-152.4	-140.9
	303	521.06	422.64	413.95	-	-	-	-	-	-
	308	518.17	419.74	411.06	-	-	-	-	-	-
	313	515.27	413.95	402.38	-	-	-	-	-	-
5083	298	581.85	480.54	466.06	578.9	578.9	578.9	-409.6	-308.2	-293.8
	303	578.96	477.64	463.17	-	-	-	-	-	-
	308	576.06	474.75	460.27	-	-	-	-	-	-
	313	573.17	471.85	457.38	-	-	-	-	-	-
7075	298	578.96	474.75	454.48	578.9	1563.2	578.9	-406.7	-8.97	-282.2
	303	576.06	466.06	451.59	-	-	-	-	-	-
	308	573.17	457.38	448.69	-	-	-	-	-	-
	313	570.27	451.59	445.80	-	-	-	-	-	-

Tables (6-9) : *The thermodynamic quantities for pure Al and its alloy in pH=11 in presence of Na₂CrO₄ at four temperatures.*

Elect.	T (K)	-ΔG (kJ.mole⁻¹)			ΔS (J.mol⁻¹.K⁻¹)			ΔH (kJ.mole⁻¹)		
		0.005	0.01	0.05	0.005	0.01	0.05	0.005	0.01	0.05
Pure Al	298	578.96	440.01	437.11	1215.8	1331.6	1736.8	-216.8	-43.37	80.21
	303	576.06	431.32	425.53	-	-	-	-	-	-
	308	567.38	425.53	416.85	-	-	-	-	-	-
	313	561.59	419.74	411.06	-	-	-	-	-	-
2024	298	532.64	422.64	550.01	578.9	810.5	1389.5	-360.3	-181.2	-136.1
	303	529.75	419.74	541.33	-	-	-	-	-	-
	308	526.85	413.95	532.64	-	-	-	-	-	-
	313	523.96	411.06	529.75	-	-	-	-	-	-
5083	298	541.33	416.85	495.01	578.9	752.6	2200.1	-369.1	-192.7	160.5
	303	538.43	411.06	492.11	-	-	-	-	-	-
	308	535.54	408.16	477.64	-	-	-	-	-	-
	313	532.64	405.27	463.17	-	-	-	-	-	-
7075	298	570.27	411.06	497.90	984.2	578.9	1505.3	-277.0	-238.8	-49.41
	303	564.48	408.16	492.11	-	-	-	-	-	-
	308	558.70	405.27	486.32	-	-	-	-	-	-
	313	555.80	402.38	474.75	-	-	-	-	-	-

Tables (6-10) : *The thermodynamic quantities for pure Al and its alloy in pH=9 in presence of Na₂CrO₄ at four temperatures.*

Elect.	T (K)	-ΔG (kJ.mole⁻¹)			ΔS (J.mol⁻¹.K⁻¹)			ΔH (kJ.mole⁻¹)		
		0.005	0.01	0.05	0.005	0.01	0.05	0.005	0.01	0.05
Pure Al	298	541.33	445.80	578.96	926.3	752.6	1157.9	-265.3	-221.7	-234.1
	303	535.54	440.01	573.17	-	-	-	-	-	-
	308	532.64	437.11	567.38	-	-	-	-	-	-
	313	526.85	434.22	561.59	-	-	-	-	-	-
2024	298	573.17	376.32	506.59	2721.1	1679	2489.5	237.6	123.7	235.1
	303	552.91	361.85	483.43	-	-	-	-	-	-
	308	538.43	356.06	471.85	-	-	-	-	-	-
	313	532.64	350.27	468.96	-	-	-	-	-	-
5083	298	526.85	402.38	518.17	578.9	810.5	3010.6	-354.6	-161.0	378.8
	303	523.96	399.48	509.48	-	-	-	-	-	-
	308	521.06	393.69	497.90	-	-	-	-	-	-
	313	518.17	390.80	471.85	-	-	-	-	-	-
7075	298	547.12	405.27	558.70	1157.9	810.5	2547.4	-202.3	-163.8	200.3
	303	541.33	402.38	552.91	-	-	-	-	-	-
	308	535.54	396.59	538.43	-	-	-	-	-	-
	313	529.75	393.69	521.06	-	-	-	-	-	-

6-1-1 Results And Discussion Of ΔG

When a metal undergoes corrosion there is a change in Gibbs free energy, ΔG, of the system, which is equal to the work, associated with the corrosion reaction.

The performance of such a work is accompanied usually by a decrease in the Gibbs free energy of system. Figs. (6-6) to (6-9) show the variation of the ($-\Delta G$) for corrosion of pure Al and its alloy in the research conditions, generally values of ΔG were negative suggesting the existence of thermodynamic feasibility for the corrosion of the electrodes materials in NaOH solution in the absence or the presence of the additives in the basic media, the results may be summarized in the following :

1- In NaOH solution without additives, in general, it is shown that the ($-\Delta G$) values are arranged in a sequence as:

$-\Delta G$ (kJ.mole^{-1})

at constant pH and   *7075 > 5083 > pure Al > 2024*

   temperature

this is means that the corrosion process of 7075 alloy in the basic media is more spontaneous than the other electrodes in the same medium.

But for the same electrode at a constant temperature it is shown :

$-\Delta G$ in term pH   *13 > 11 > 9*

This means that the increase of the concentration of medium (NaOH solution) shift the ΔG to more negative values and suggesting spontaneously of the process, as shown in Fig. (6-6).

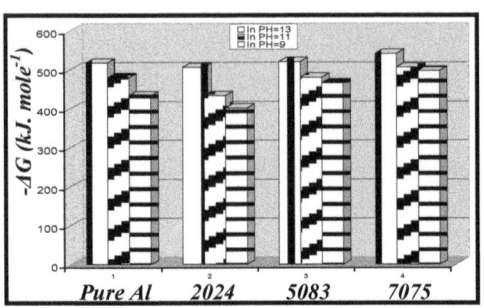

Fig. (6-6) : *Values of (-ΔG)plotted for*
pure Al and its alloys in NaOH solution
in the absence of additives at 298K.

2- In the presence of sodium chloride in NaOH solution, it is shown that -ΔG values may be arranged in the following sequence, as shown in Fig. (6-7 a, b, c)

-ΔG (kJ. mole⁻¹)

   at constant pH,   *pure Al > 7075 > 5083 > 2024*

[Cl⁻], and Temp.

This means that the presence of chloride ions causes increasing of corrosion process for pure Al more than alloys and makes this process more spontaneous. For the same electrode at a constant temperature, -ΔG values vary with variation of concentration of NaCl additive as:

-ΔG in term [Cl⁻]

   In pH=13   *$10^{-3} > 10^{-2} > 0.1$ mol.dm⁻³*

While in pH=11   *$10^{-2} > 0.1 > 10^{-3}$ mol.dm⁻³*

Generally, in pH=9, there are various behaviour of –ΔG for pure Al and its alloys as shown in Fig. (6-7 c).

3- In the presence of sodium acetate as an organic inhibitor in NaOH solution, it is shown that $-\Delta G$ values may be arranged in the following sequence, as shown in Fig. (6-8 a, b, c) :

$-\Delta G$ (kJ. mole^{-1})

at constant pH, Temp.,   ***pure Al > 7075 > 5083 > 2024***

and [CH$_3$COO$^-$]

 This means that the presence of sodium acetate also makes corrosion process for pure Al more spontaneous than alloys in three values of pH in the solution.

 But for same electrode vary $(-\Delta G)$ values with variation of concentration of sodium acetate.

4- In the presence of sodium chromate as inorganic inhibitor in NaOH solution, it is shown that $-\Delta G$ values may be arranged in the following sequence, as shown in Fig. (6-9 a, b, c) :

$-\Delta G$ in pH=13

at constant Temp.   ***pure Al > 5083 > 7075 > 2024***

and [CrO$_4^{2-}$]

$-\Delta G$ in pH=9

at constant Temp.   ***pure Al > 7075 > 5083 > 2024***

and [CrO$_4^{2-}$]

while a different behaviour is observed in pH=11with different the concentrations of sodium chromate, at a constant temperature, as shown below:

	$5x10^{-3}$ mol.dm^{-3} CrO$_4^{-2}$	***pure Al > 7075 > 5083 > 2024***
$-\Delta G$	*$1x10^{-2}$ mol.dm^{-3} CrO$_4^{-2}$*	***pure Al > 2024 > 5083 > 7075***
	$5x10^{-2}$ mol.dm^{-3} CrO$_4^{-2}$	***2024 > 7075 > 5083 > pure Al***

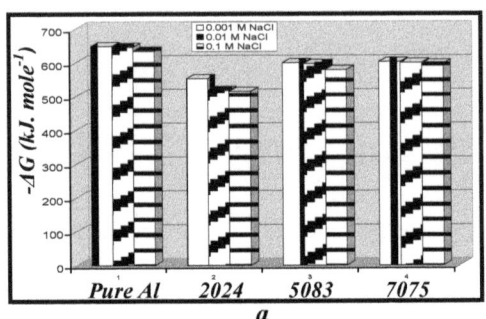

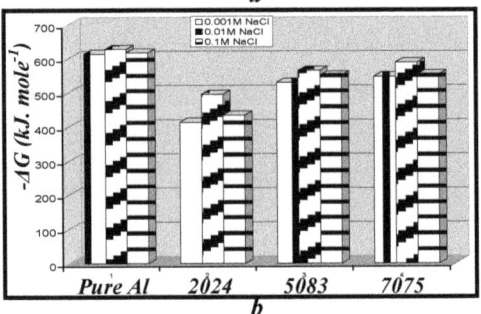

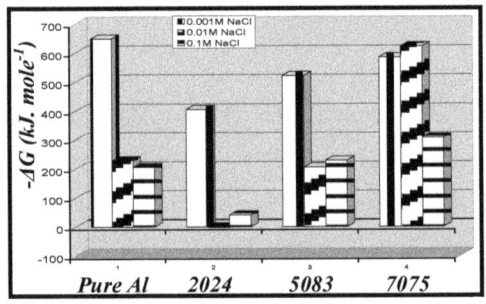

c

Fig. (6-7) : *Values of (-ΔG)plotted for*
pure Al and its alloys in NaOH solution
in the presence of NaCl at 298K.
a: in pH=13, b: in pH=11, c: in pH=9

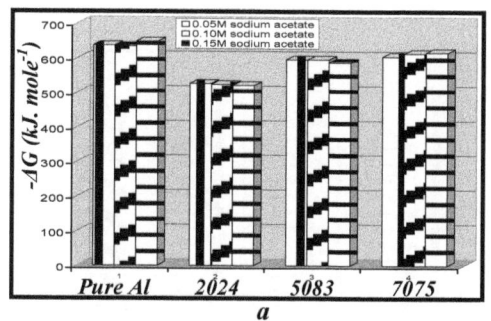

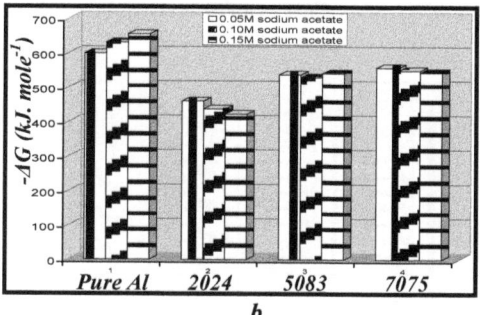

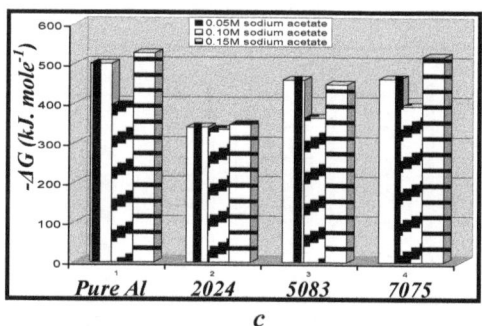

Fig. (6-8) : *Values of (-ΔG)plotted for*
pure Al and its alloys in NaOH solution
in the presence of CH₃COONa at 298K.
a: in pH=13, b: in pH=11, c: in pH=9

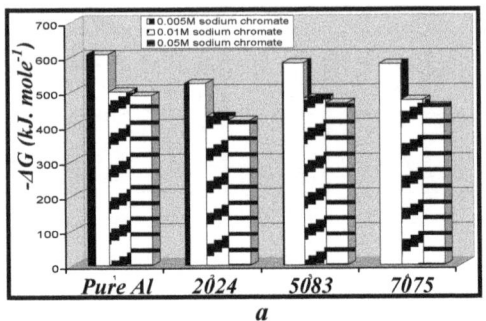

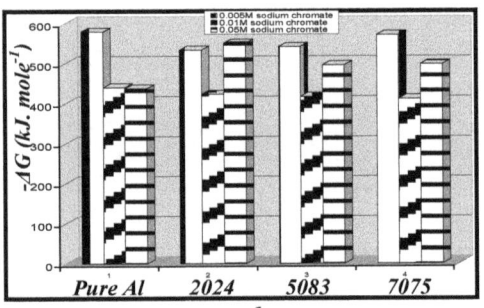

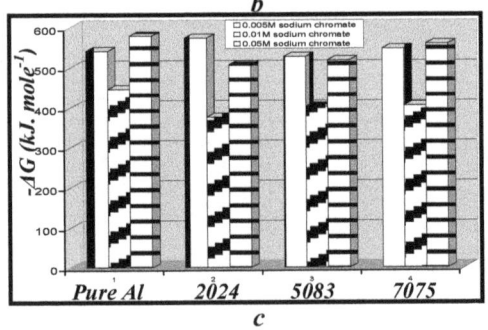

Fig. (6-9) : _Values of (-ΔG) plotted for_
pure Al and its alloys in NaOH solution
in the presence of Na₂CrO₄ at 298K.
_a: in pH=13, b: in pH=11, c: in pH=9_

206

6-1-2 Results And Discussion Of ΔS

Values of (ΔS) were positive or negative depending on the positive or negative dependencies of (ΔG) values on temperatures. Values of (ΔS) reflect the change in the order and orientation of the solvent molecules around the hydrated metal ions in the corrosion medium when metal atoms were corroded and subsequently hydrated in the solution.

Values of ΔS were generally positive due to negativity of ΔG, this suggests a lower order in the solvated states of the metal ions as compared with the state of metal atoms in the crystal lattice of the corroding electrodes. Results of Tables (6-1) to (6-10) indicate the variation of (ΔS) with increasing the concentration of NaOH solution in the absence and presence of additives. Fig. (6-10) show the variation of (ΔS) for corrosion of pure Al and its alloys in NaOH solution with three pH values in the absence of additives.

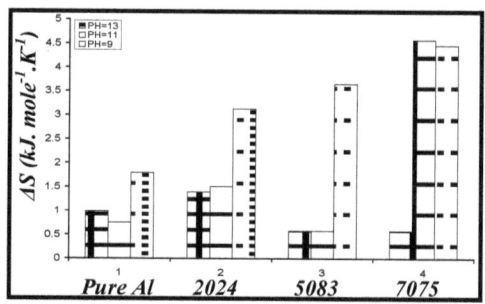

Fig. (6-10) : *Values of (ΔS) plotted for*
pure Al and its alloys in NaOH solution
in the absence of additives.

6-1-3 Results And Discussion Of ΔH

Values of the enthalpy of corrosion (ΔH) reflect the enthalpy changes associated with the corrosion reaction and ranged from negative to positive values indicating exothermic or endothermic nature of corrosion reaction. The value of (ΔH) may be estimated from such relation as:

$$\Delta H = S + I_M - W_{M^{z+}} - (\frac{1}{2}D_{H_2} + I_H - W_{H^+}) \quad \ldots\ldots(6\text{-}4)$$

where:

S: the heat of sublimation of the metal M,

I_M: is the ionization potential of the metal M,

$W_{M^{z+}}$: is the heat hydration of M^{z+} ions in the solution,

D_{H_2} : is the bond – dissociation energy of the hydrogen molecules,

I_H: is the ionization potential of hydrogen atoms,

W_{H^+} : is the heat of hydrogen of hydrogen ions.

When comparing the enthalpy changes accompanying various metals in the same medium, the factors affecting the values of (ΔH) should be confined to :

$$S + I_M - W_{M^{z+}} \quad \ldots(6\text{-}5)$$

These three quantities play an important role in deciding the extent of variation in (ΔH) values for the metal and for its alloys in the same medium.

The results of Tables (6-1) to (6-10) indicates that , generally, (ΔH) values were negative in pH=13 for pure Al and its alloys, as shown in Fig. (6-11*a*), and arranged in the following sequence:

(-ΔH) value   *7075 > 5083 > pure Al > 2024*

208

While (ΔH) values were positive in pH=9 for pure Al and its alloys, as shown in Fig. (6-11 c) and arranged in the following sequence:

($+\Delta H$) value $7075 > 5083 > 2024 > pure\ Al$

But in pH=11, (ΔH) values were negative for pure Al and 5083 alloy, and positive for 2024 and 7075 alloy, as shown in Fig. (6-11 b) and arranged in the following sequence:

(ΔH) value $7075 > 2024 > pure\ Al > 5083$

In the presence of NaCl in solution, (ΔH) were negative for pure Al, 5083, and 7075 alloy while positive for 2024 alloy in pH=13 and 11. But in pH=9, (ΔH) values were positive for pure Al and its alloys.

In presence of CH_3COONa in solution, (ΔH) values were negative for pure Al, 5083, and 7075 alloy, while positive for 2024 alloy in PH=13 and 11. But in PH=9, there are various behaviour of (ΔH).

In the presence of Na_2CrO_4 , (ΔH) values were negative for pure Al and its alloys in pH=13, but (ΔH) values were negative for 2024 and 7075 alloy and takes various behaviour for pure Al and 5083 alloy in pH=11.

While in pH=9, (ΔH) values were negative for pure Al, positive for 2024 alloy, and different for 5083 and 7075 alloys.

Generally, negative values of (ΔH) indicating a stronger bonding of the metal ions, resulting from electrode corrosion, with the species that are present in the corrosion medium as compared with the bonding of the metal atoms in the crystal lattice of the solid electrode.

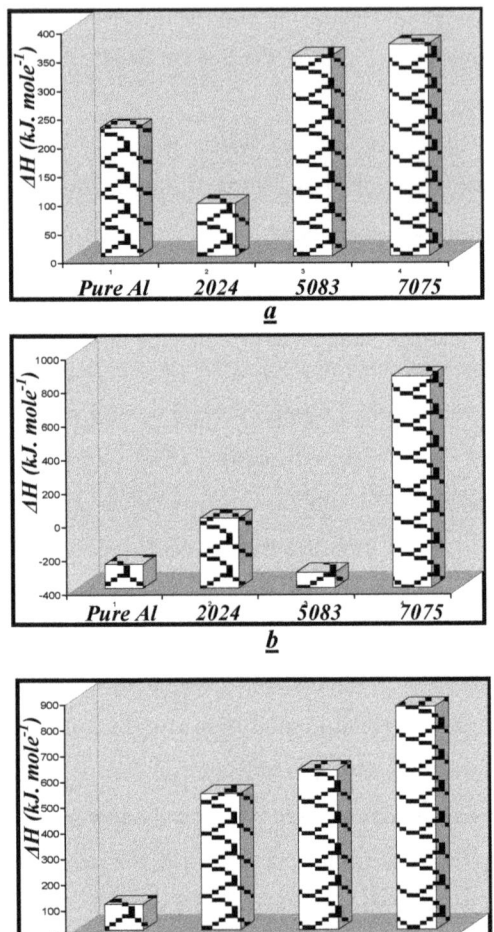

Fig. (6-11) : *Values of (ΔH) plotted for*

pure Al and its alloys in NaOH solution

in the absence of additives.

a: pH=13, b: pH=11, c: pH=9.

6-2 Kinetic Of Corrosion

The rate (r) of corrosion in a given environment is directly proportional with its corrosion current density (i_{corr}) in accordance with the relation[170, 171] :

$$r = 0.13(e/\rho)i_{corr} \qquad \ldots\ldots\ldots(6\text{-}6)$$

here (e) is the equivalent weight of the metal and (ρ) is its density. For the increasing values of (i_{corr}) with a temperature follow Arrhenius equation, it is reasonable as:

$$i_{corr} = A\exp(-E_a / RT) \qquad \ldots\ldots(6\text{-}7)$$

where A and E_a are the pre- exponential factor and energy of activation of the corrosion process respectively.

Values of E_a were derived from the slopes of the (log i_{corr}) versus (1/T) linear plots as in Fig. (6-12 a, b, c), while those of (A) were obtained from the intercepts of the plots at (1/T=zero); values of (A), expressed in term of (Amper m^{-2}), have then been converted into (molecules per m^2 per second), (A) was defined as:

$$A = \frac{KT}{h}e^{\Delta S^* / R} \qquad \ldots\ldots(6\text{-}8)$$

where

K: Boltzman constant,

h: Plank constant,

T: temperature on Kelvin scale,

R: gas constant, and

ΔS^*: the entropy of activation.

Tables (6-11) to (6-20) show the results of activation energy (E_a), log A, pre- exponential factors (A), and entropy of activation (ΔS^*) for pure Al and its alloys in NaOH solution with three values of pH (13, 11, and 9) in the absence and presence of additives.

211

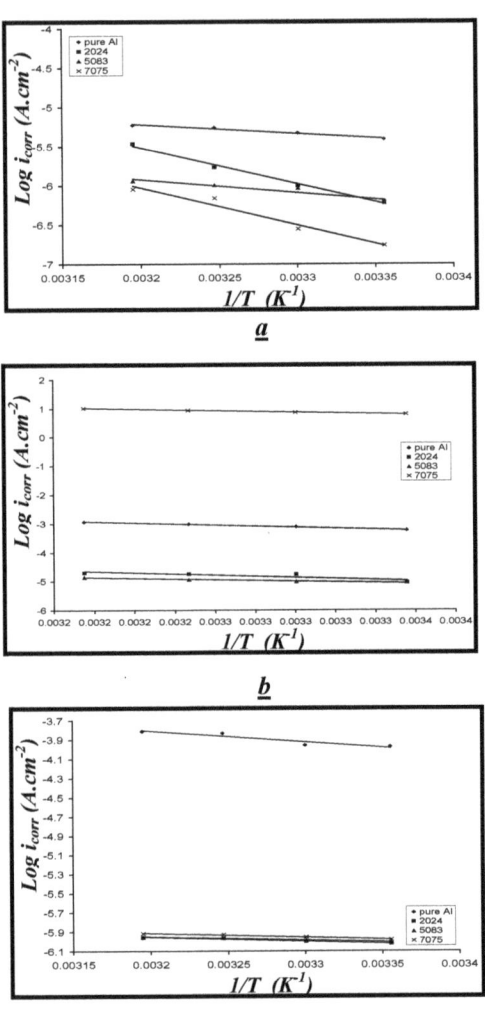

Fig. (6-12) *: Arrhenius plots, relating(log i$_{corr}$)values to (1/T) for the*
corrosion of pure Al and its alloys in NaOH solution
in the absence of additives.
a: pH=13, b: pH=11, c: pH=9.

Table (6-11) : *Values of activation energy (E_a), pre- exponential factors (A), and entropy of activation (ΔS^*) for pure Al and its alloys in NaOH solution with three values of pH (13, 11, and 9)in the absence additives.*

Elect.	E_a (kJ.mol^{-1})			Log A			A molecules.m^{-2}.s^{-1}			ΔS^* (J.mol^{-1}.K^{-1})		
	13	11	9	13	11	9	13	11	9	13	11	9
Pure Al	21.75	35.77	21.68	-1.59	3.03	-0.17	2.45×10^{20}	5.73×10^{15}	9.27×10^{18}	-119.62	-81.111	-107.79
2024	88.31	38.29	8.030	9.27	1.65	-4.61	3.34×10^{9}	1.37×10^{17}	2.57×10^{23}	-29.28	-92.564	-144.72
5083	31.87	23.86	7.100	-0.63	0.87	4.76	2.69×10^{19}	4.67×10^{19}	3.61×10^{23}	-111.64	-113.63	-145.96
7075	90.82	26.43	8.010	9.15	-0.57	-4.57	4.34×10^{9}	2.33×10^{19}	2.36×10^{23}	-30.225	-111.12	-144.42

Table (6-12) : *Values of activation energy (E_a), pre- exponential factors (A), and entropy of activation (ΔS^*) for pure Al and its alloys in pH=13 in the presence of NaCl with three concentration (10^{-3}, 10^{-2}, 0.1 mol.dm^{-3}).*

Elect.	E_a (kJ.mol^{-1})			Log A			A molecules.m^{-2}.s^{-1}			ΔS^* (J.mol^{-1}.K^{-1})		
	0.001	0.01	0.1	0.001	0.01	0.1	0.001	0.01	0.1	0.001	0.01	0.1
Pure Al	37.10	60.84	39.18	2.99	7.45	3.68	6.26×10^{15}	2.16×10^{11}	1.30×10^{15}	-81.43	-44.34	-75.75
2024	44.40	59.68	38.03	3.87	6.76	3.39	8.38×10^{14}	1.07×10^{12}	2.48×10^{15}	-74.17	-50.12	-78.09
5083	66.52	46.78	42.71	7.92	4.61	4.26	7.44×10^{10}	1.51×10^{14}	3.36×10^{14}	-40.48	-67.99	-70.87
7075	127.97	19.01	28.17	17.42	-0.36	1.47	23.28	1.45×10^{19}	2.08×10^{17}	38.53	-109.41	-94.08

Table (6-13) : *Values of activation energy (E_a), pre- exponential factors (A), and entropy of activation (ΔS^*) for pure Al and its alloys in pH=11 in the presence of NaCl with three concentration (10^{-3}, 10^{-2}, 0.1 mol.dm^{-3}).*

Elect.	E_a (kJ.mol^{-1})			Log A			A molecules.m^{-2}.s^{-1}			ΔS^* (J.mol^{-1}.K^{-1})		
	0.001	0.01	0.1	0.001	0.01	0.1	0.001	0.01	0.1	0.001	0.01	0.1
Pure Al	36.50	32.05	50.61	-0.93	-0.41	4.31	5.39×10^{19}	1.62×10^{19}	3.02×10^{14}	-114.14	-109.81	-70.49
2024	34.88	20.98	55.40	0.92	-2.15	4.88	7.48×10^{17}	8.84×10^{20}	8.21×10^{13}	-98.70	-124.24	-65.78
5083	101.57	37.29	18.71	11.76	0.54	-3.13	10.71×10^{-6}	1.79×10^{18}	8.58×10^{21}	-8.54	-101.86	-132.45
7075	52.37	21.73	54.06	2.08	-2.31	2.85	5.10×10^{16}	1.27×10^{21}	8.75×10^{15}	-89.00	-125.56	-82.63

Table (6-14) : *Values of activation energy (E_a), pre- exponential factors (A), and entropy of activation (ΔS^*) for pure Al and its alloys in pH=9 in the presence of NaCl with three concentration (10^{-3}, 10^{-2}, 0.1 mol.dm^{-3}).*

Elect.	E_a (kJ.mol^{-1})			Log A			A molecules.m^{-2}.s^{-1}			ΔS^* (J.mol^{-1}.K^{-1})		
	0.001	0.01	0.1	0.001	0.01	0.1	0.001	0.01	0.1	0.001	0.01	0.1
Pure Al	32.40	4.49	32.14	1.35	4.42	-3.58	1.05×10^{19}	2.41×10^{25}	3.61×10^{20}	-108.27	-161.13	-121.02
2024	12.48	19.65	69.10	-3.81	-3.65	5.00	4.10×10^{22}	2.84×10^{22}	6.13×10^{13}	-138.10	-136.78	-64.73
5083	10.69	4.92	30.14	-3.78	-6.29	-1.79	8.31×10^{16}	1.23×10^{25}	3.92×10^{20}	-90.77	-158.71	-121.31
7075	47.43	64.36	25.67	-0.22	-6.58	-1.76	2.74×10^{17}	2.36×10^{14}	2.38×10^{22}	-95.67	-69.60	-136.13

Table (6-15) : *Values of activation energy (E_a), pre- exponential factors (A), and entropy of activation (ΔS^*) for pure Al and its alloys in pH=13 in the presence of CH_3COONa with three concentration (0.05, 0.1, 0.15 mol.dm^{-3}).*

Elect.	E_a (kJ.mol^{-1})			Log A			A molecules.m^{-2}.s^{-1}			ΔS^* (J.mol^{-1}.K^{-1})		
	0.05	0.10	0.15	0.05	0.10	0.15	0.05	0.10	0.15	0.05	0.10	0.15
Pure Al	36.44	50.86	16.13	3.02	5.30	-0.52	5.91×10^{15}	3.09×10^{13}	2.09×10^{19}	-81.22	-62.25	-110.74
2024	26.98	32.02	28.16	1.41	2.15	1.59	2.38×10^{17}	4.41×10^{16}	1.58×10^{17}	-94.57	-88.48	-93.10
5083	15.67	22.81	40.10	-0.55	0.77	3.70	2.21×10^{19}	1.04×10^{18}	1.23×10^{15}	-110.93	-99.91	-75.55
7075	19.92	81.50	12.58	-0.01	10.45	-1.15	6.47×10^{18}	2.19×10^{8}	8.87×10^{19}	-106.49	-19.43	-115.94

Table (6-16) : *Values of activation energy (E_a), pre- exponential factors (A), and entropy of activation (ΔS^*) for pure Al and its alloys in pH=11 in the presence of CH_3COONa with three concentration (0.05, 0.1, 0.15 mol.dm^{-3}).*

Elect.	E_a (kJ.mol^{-1})			Log A			A molecules.m^{-2}.s^{-1}			ΔS^* (J.mol^{-1}.K^{-1})		
	0.05	0.10	0.15	0.05	0.10	0.15	0.05	0.10	0.15	0.05	0.10	0.15
Pure Al	35.65	39.34	47.74	1.16	1.99	3.52	4.22×10^{17}	6.35×10^{16}	1.86×10^{15}	-96.63	-89.79	-7.05
2024	48.59	40.06	43.48	3.17	2.07	2.76	4.13×10^{15}	5.24×10^{16}	1.06×10^{16}	-79.93	-89.10	-83.34
5083	7.38	31.85	39.86	-3.39	0.94	2.39	1.53×10^{22}	7.09×10^{17}	2.54×10^{16}	-134.55	-98.51	-86.48
7075	26.81	45.18	32.56	-0.28	3.11	1.09	1.20×10^{19}	4.82×10^{15}	5.04×10^{17}	-108.71	-80.48	-97.27

Table (6-17) : *Values of activation energy (E$_a$), pre- exponential factors (A), and entropy of activation (ΔS*) for pure Al and its alloys in pH=9 in the presence of CH$_3$COONa with three concentration (0.05, 0.1, 0.15 mol.dm^{-3}).*

Elect.	E$_a$ (kJ.mol^{-1})			Log A			A molecules.m^{-2}.s^{-1}			ΔS* (J.mol^{-1}.K^{-1})		
	0.05	0.10	0.15	0.05	0.10	0.15	0.05	0.10	0.15	0.05	0.10	0.15
Pure Al	19.30	44.61	23.46	-3.98	0.53	-3.10	5.97×10^{22}	1.80×10^{18}	8.02×10^{21}	-139.45	-101.87	-132.21
2024	22.63	21.40	22.38	-3.30	-3.25	-3.28	1.24×10^{22}	1.12×10^{22}	1.19×10^{22}	-133.80	-133.42	-133.66
5083	17.81	19.00	20.82	-4.10	-3.48	-3.54	7.87×10^{22}	1.91×10^{22}	2.17×10^{22}	-140.45	-135.35	-135.80
7075	20.79	27.99	22.02	-3.59	-2.07	-3.33	2.44×10^{22}	7.44×10^{20}	1.36×10^{22}	-136.22	-123.62	-134.12

Table (6-18) : *Values of activation energy (E$_a$), pre- exponential factors (A), and entropy of activation (ΔS*) for pure Al and its alloys in pH=13 in the presence of Na$_2$CrO$_4$ with three concentration(0.005, 0.01, 0.05 mol.dm^{-3}).*

Elect.	E$_a$ (kJ.mol^{-1})			Log A			A molecules.m^{-2}.s^{-1}			ΔS* (J.mol^{-1}.K^{-1})		
	0.005	0.01	0.05	0.005	0.01	0.05	0.005	0.01	0.05	0.005	0.01	0.05
Pure Al	21.32	21.47	18.07	0.49	0.62	0.07	1.99×10^{18}	1.49×10^{18}	5.18×10^{18}	-102.25	-101.19	-105.69
2024	20.88	12.66	11.93	0.43	-0.83	-0.92	2.28×10^{18}	4.30×10^{19}	5.23×10^{19}	-102.72	-113.33	-114.04
5083	13.22	8.90	13.65	-0.94	-1.58	-0.72	5.44×10^{19}	2.40×10^{20}	3.32×10^{19}	-114.18	-119.55	-112.40
7075	21.30	39.44	27.16	0.26	3.23	1.04	3.42×10^{18}	3.67×10^{15}	5.58×10^{17}	-104.19	79.50	-97.64

Table (6-19) : *Values of activation energy (E_a), pre- exponential factors (A), and entropy of activation (ΔS^*) for pure Al and its alloys in pH=11 in the presence of Na_2CrO_4 with three concentration (0.005, 0.01, 0.05 mol.dm^{-3}).*

Elect.	E_a (kJ.mol^{-1})			Log A			A molecules.m^{-2}.s^{-1}			ΔS^* (J.mol^{-1}.K^{-1})		
	0.005	0.01	0.05	0.005	0.01	0.05	0.005	0.01	0.05	0.005	0.01	0.05
Pure Al	38.36	51.22	18.71	-0.55	1.93	-3.80	2.25×10^{19}	7.16×10^{16}	4.01×10^{22}	-111.00	-90.23	-138.02
2024	18.75	12.96	10.74	-3.65	-4.17	-4.64	2.80×10^{22}	9.27×10^{22}	2.73×10^{23}	-136.73	-141.05	-144.95
5083	9.21	27.76	14.36	-5.43	-1.83	-4.31	1.69×10^{24}	4.23×10^{20}	1.28×10^{23}	-151.53	-121.59	-142.22
7075	23.04	22.31	11.75	-3.43	-3.20	-5.11	1.69×10^{22}	9.90×10^{21}	8.07×10^{23}	-134.91	-132.97	-148.86

Table (6-20) : *Values of activation energy (E_a), pre- exponential factors (A), and entropy of activation (ΔS^*) for pure Al and its alloys in pH=9 in the presence of Na_2CrO_4 with three concentration (0.005, 0.01, 0.05 mol.dm^{-3}).*

Elect.	E_a (kJ.mol^{-1})			Log A			A molecules.m^{-2}.s^{-1}			ΔS^* (J.mol^{-1}.K^{-1})		
	0.005	0.01	0.05	0.005	0.01	0.05	0.005	0.01	0.05	0.005	0.01	0.05
Pure Al	13.33	39.24	14.91	-4.05	0.17	-4.58	7.10×10^{22}	4.21×10^{18}	2.30×10^{23}	-140.08	-104.94	-144.47
2024	29.36	15.90	21.92	-1.23	-3.65	-3.31	1.07×10^{20}	2.84×10^{22}	1.28×10^{22}	-116.63	-136.78	-133.91
5083	20.65	23.56	22.47	-2.61	-2.30	-3.21	2.58×10^{21}	1.25×10^{21}	1.02×10^{22}	-128.11	-125.52	-13.07
7075	16.81	15.99	21.81	-3.43	-3.71	-3.37	1.71×10^{22}	3.24×10^{22}	1.46×10^{22}	-134.95	-137.25	-134.38

Tables (6-11) to (6-20) show the resulting values of (E_a) and $(\log A)$ for the corrosion of pure aluminium and its alloys, the results of (E_a) may be summarized in the following :

1- The (E_a) values in NaOH solution in the absence of additives may be arranged in a sequence as [Fig. (6-13)]:

In pH=13   *(E_a)*   *7075 > 2024 > 5083 > pure Al*

In pH=11   *(E_a)*   *2024 > pure Al > 7075 > 5083*

In pH=9   *(E_a)*   *pure Al > 2024 $\simeq$ 7075 > 5083*

2- Generally, the presence of additives in NaOH solution gives different values of (E_a) according to the type of additive and its concentration.

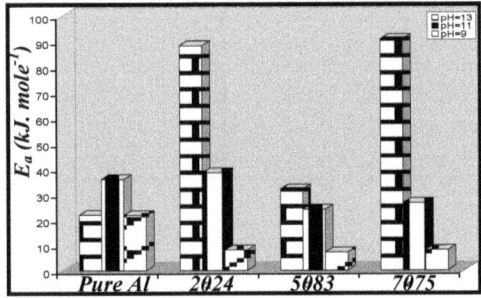

***Fig. (6-13)* :** *Values of (E_a) plotted for*
pure Al and its alloys in NaOH solution
in the absence of additives.

The $(\log A)$ values (ranged between positive and negative values) which were summarized in the following :

1- In NaOH solution in the absence of additives as shown in Fig. (6-14), may be arranged in the following sequence :

In pH=13 (log A)   *2024 $\simeq$ 7075 > 5083 > pure Al*

In pH=11 (log A)   *pure Al > 2024 > 5083 > 7075*

In pH=9   (log A)   *5083 > pure Al > 7075 $\simeq$ 2024*

2- Generally, in the presence of additives, (log A) values may be arranged in a sequence as (except some cases) :

(log A) for pure Al and

its alloys in NaOH solution $NaCl > CH_3COONa > Na_2CrO_4$

in presence of additives

The highest value of activation energy was found (127.97 kJ.mol^{-1}) for 7075 alloy and the highest value of **log A** was found (17.42) in pH=13 in presence of (1x10^{-3} mol.dm^{-3}) NaCl, while the lowest value of $\mathbf{E_a}$ and **log A** was found 4.49 for pure Al in pH=9 + 1x10^{-2} mol.dm^{-3} NaCl) and -6.58 (for 7075 alloy in pH=9 + 1x10^{-2} mol.dm^{-3} NaCl) respectively.

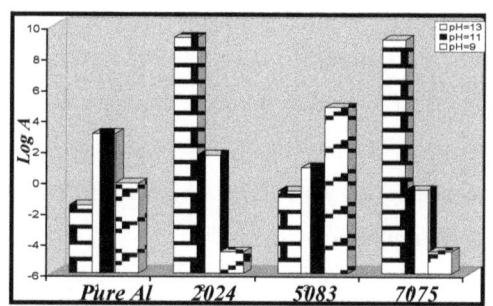

Fig. (6-14) : Values of (Log A) plotted for pure

Al and its alloys in NaOH solution in the

absence of additives.

The relationship existed between values of the activation energy (E_a) and logarithm of pre – exponential factor (log A) in different media suggesting the operation of a compensation effect in kinetics of corrosion. This suggests that, the corrosion reaction proceeds on surface sites, which were associated with different energies of activation. The corrosion reaction is assumed to start on sites with lower (E_a) and (log A) values first, spending thereafter to those sites on which (Ea) and (log A) were higher.

The results of Figs. (6-15) to (6-18) indicate the existence of a linear relationship between the values of (log A) and the corresponding values of (E_a) which may be expressed as[172] :

$$\log A = mE_a + I \qquad \qquad ......(6\text{-}9)$$

where m and I are respectively the slope and intercept of the plots, such a behaviour is referred to as "compensation effect" which describes the kinetics of a great number of catalytic and tarnishing reactions on metals[173, 174].

Equation (6-9) indicates that simultaneous increases or decreases in (E_a) and (log A) for a system tend to compensate from the standpoint of the reaction rate.

A number of interpretations[175] have been offered for the phenomenon of the compensation effect in surface reaction, among which the effect could be ascribed to the presence of energetically heterogeneous reaction solution. A decrease in (E_a) at constant (log A) implies a higher rate, while an increase in (E_a) at constant (log A) implies a lower rate.

Simultaneous increase in (E_a) and (log A) therefore tend to compensate from the standpoint of the corrosion rate. When such a compensation operates, it is possible for striking variations in (E_a) and (log A) through a series of surface sites on a metal or an alloy to yield only a small variation in reactivity.

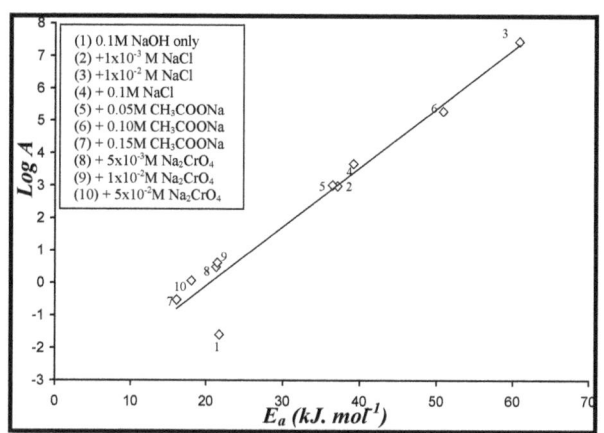

Fig. (6-15) : *Log A values plotted against E_a for pure Al in 0.1mol.dm^{-3} NaOH solution in the absence and presence of additives.*

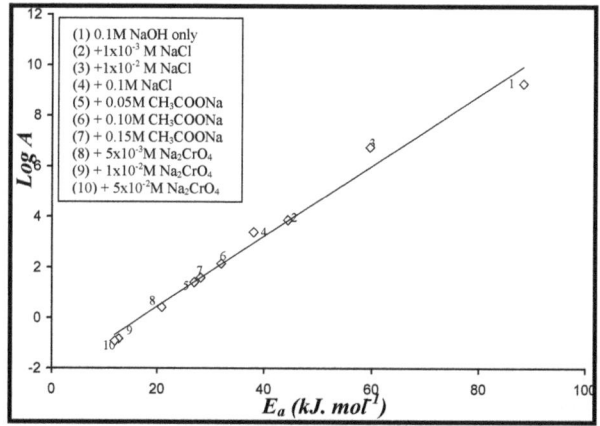

Fig. (6-16) : *Log A values plotted against E_a for 2024 alloy in 0.1mol.dm^{-3} NaOH solution in the absence and presence of additives.*

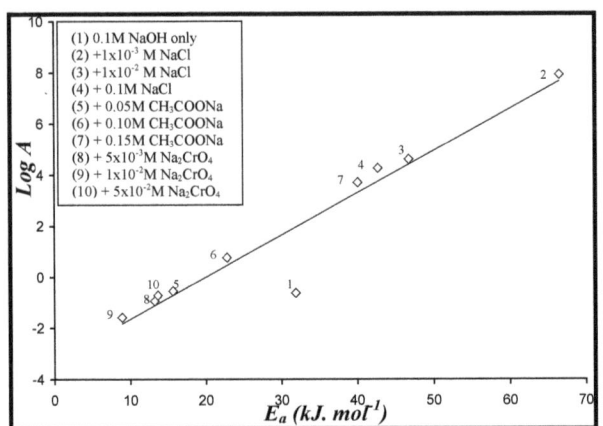

Fig. (6-17) : *Log A values plotted against E*$_a$ *for*
5083 alloy in 0.1mol.dm^{-3} *NaOH solution in the*
absence and presence of additives.

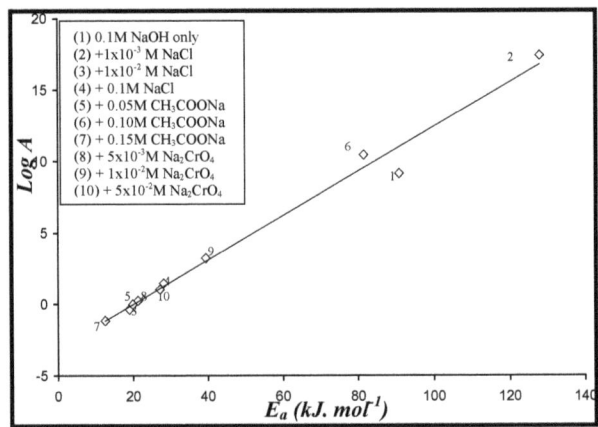

Fig. (6-18) : *Log A values plotted against E*$_a$ *for*
7075 alloy in 0.1mol.dm^{-3} *NaOH solution in the*
absence and presence of additives.

Finally, ΔS^* values in the presence of different amounts of additives shifted to more or less negative values than the corresponding values in the absence of additives indicating a decrease in the rate corrosion of electrodes. Fig. (6-19) show the variation of (ΔS^*) for pure Al and its alloys in NaOH solution, the results may be arranged in a sequence as :

In pH=13   *pure Al > 5083 > 7075 > 2024*

(-ΔS*)   *In pH=11*   *5083 > 7075 > 2024 > pure Al*

    In pH=9   *5083 > 2024 ≥ 7075 > pure Al*

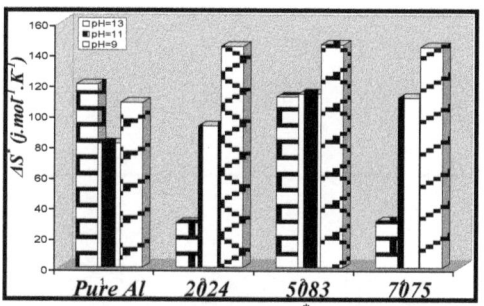

Fig. (6-19) : *Values of (ΔS*) plotted for pure Al and its alloys in NaOH solution in the absence of additives.*

Generally, in the presence of additives, (ΔS^*) values may be arranged in a following sequence (except some cases), as shown in Fig. (6-20) to (6-23) for pure Al and 2024, 5083, and 7075 alloy in pH=13 respectively:

(ΔS*) for pure Al and

its alloys in NaOH solution   *NaCl > CH₃COONa > Na₂CrO₄*

in presence of additives

223

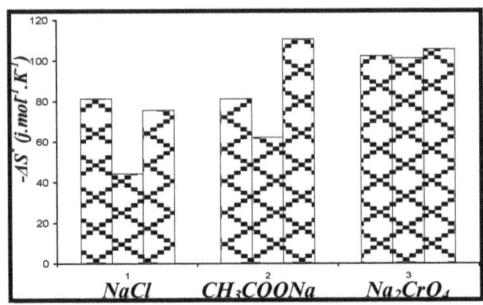

Fig. (6-20) *: Values of (ΔS*) plotted for*
pure Al in pH=13 in the presence
of additives.

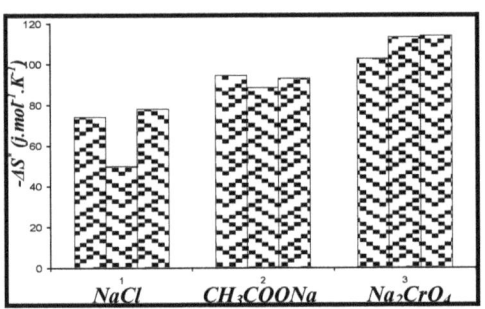

Fig. (6-21) *: Values of (ΔS*) plotted for 2024*
alloy in pH=13 in the presence
of additives.

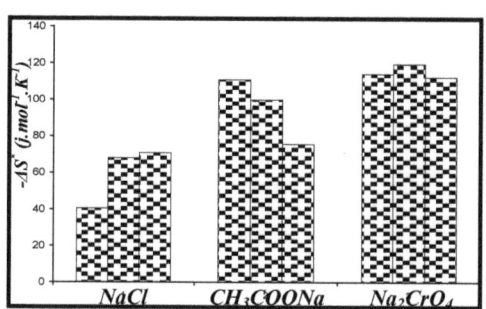

Fig. (6-22) *: Values of (ΔS^{*}) plotted for 5083*

alloy in pH=13 in the presence

of additives.

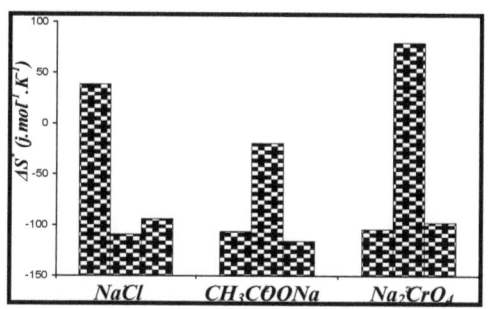

Fig. (6-23) *: Values of (ΔS^{*}) plotted for 7075*

alloy in pH=13 in the presence

of additives.

Chapter Seven: Conclusions

7-1 Conclusions

The electrochemical behaviour of pure aluminium and of three of its alloys has been investigated in the basic media (with three values of pH) over temperatures from 298 to 313K. The effect of certain additives on the behaviour has also been examined where included sodium chloride, sodium acetate, and sodium chromate.

The study also covered the kinetic and thermodynamic aspects of the electrochemical behaviour. The conclusions that could be drawn from the experimental results and the related discussions may be put as :

I: Polarization behaviour study:

1- Polarization Studies in the basic media :

The corrosion resistance for pure aluminium and its alloys in the basic media with three values of pH may be summarized as in the following order in terms of increasing corrosion potential (decreasing corrosion) at constant pH and temperature :

Increasing (E_{corr})

[Increasing the corrosion   2024 alloy > pure Al > 5083 alloy > 7075 alloy

   resistance]   Al-Cu-Mg   99.99% Al   Al-Mg   Al-Zn-Mg

At a constant temperature, for the same electrode (pure Al, 2024, 5083, or 7075), corrosion increases with increasing concentration of the medium (NaOH solution) or increasing pH value toward basicity and the following order indicates this behaviour in term corrosion potential:

Increasing (-E_{corr})

[increasing corrosion   pH=13 > pH=11 > pH=9

in term pH value]

2- In the presence of chloride ions in solution of NaOH:

The halide ions are adsorbed on the passive film and chemically bonded with (Al^{3+}) in the passive film lattice. After dissolution the halide ions form a transitional complexes which rapidly undergo hydrolysis.

In pH=13 at a constant concentration of NaCl and temperature, (E_{corr}) become more noble towards the right as given in the following sequences at three concentration of NaCl :

(E_{corr}) values   *pure Al,   7075,   5083,   2024*

Noble Direction →

While at a constant concentration of NaCl and temperature (i_{corr}) values takes the following sequences:

(i_{corr}) values   *pure Al > 5083 alloy > 2024 alloy > 7075 alloy*

But to the same electrode at a constant temperature, increasing of the concentration of NaCl shift the (E_{corr}) toward the noble direction for pure Al and its alloys, and (i_{corr}) values increases.

In pH = 11 at constant concentration of NaCl and temperature, (E_{corr}) become more noble towards the right and more active towards the left, as given in the following sequences at three concentration of NaCl :

(E_{corr}) values   *pure Al,   7075,   5083,   2024*

Noble Direction →

While (i_{corr}) varies with variation of the concentration for pure Al and its alloy, this is indicates sensitivity of pure Al to chloride ions in pH =11.

In pH = 9, (E_{corr}) and (i_{corr}) gives different behaviour for pure Al and its alloys in this value of pH (because of very small concentration of medium), also there are no certain sequence in this case neither at a constant concentration of NaCl nor at a constant temperature.

228

Generally, increases of temperature shift the (E_{corr}) toward to noble direction except several cases, also increases of temperature lead to increasing of (i_{corr}) in all cases.

II: Inhibition of corrosion in the basic media:

In the limit of this work, it was observed that the best inhibition by using sodium acetate as an organic inhibitor gets at pH=9 with 0.05 mol.dm^{-3} of this inhibitor, and according to the sequence:

$$P\% \qquad 2024 > pure\ Al > 7075 > 5083$$

While the best inhibition by using sodium chromate as inorganic inhibitor get at pH=11 with 5×10^{-3} mol.dm^{-3} of this inhibitor for pure Al and its alloys.

III: Kinetic and thermodynamic of corrosion in the basic media:

1- In NaOH solution without additives, it is shown that the ($-\Delta G$) values may be arranged in a sequence as:

$-\Delta G\ (kJ.mole^{-1})$

at constant pH and $7075 > 5083 > pure\ Al > 2024$

   temperature

But for the same electrode at a constant temperature it is shown :

$-\Delta G\ in\ term\ pH$ $13 > 11 > 9$

In the presence of sodium chloride in NaOH solution, it is shown that ($-\Delta G$) values are arranged in the following sequence:

$-\Delta G\ (kJ.\ mole^{-1})$

 at constant pH, $pure\ Al > 7075 > 5083 > 2024$

$[Cl^{-}]$, *and Temp.*

This means that the presence of chloride ions causes increasing of corrosion process for pure Al more than alloys and makes this process more spontaneous. For the same electrode at a constant temperature, -ΔG values vary with variation of concentration of NaCl additive as:

-ΔG in term [Cl⁻]

   In pH=13 $10^{-3} > 10^{-2} > 0.1 \ mol.dm^{-3}$

While in pH=11 $10^{-2} > 0.1 > 10^{-3} \ mol.dm^{-3}$

Generally, in pH=9, there are various behaviour of –ΔG for pure Al and its alloys.

In the presence of sodium acetate as an organic inhibitor in NaOH solution, it is shown that (–ΔG) values may be arranged in the following sequence:

  -ΔG (kJ. mole⁻¹)

at constant pH, Temp.,   *pure Al > 7075 > 5083 > 2024*

   and [CH₃COO⁻]

This means that the presence of sodium acetate makes the corrosion process for pure Al more spontaneous than alloys in three values of pH in the solution. But for the same electrode (–ΔG) values vary with variation of concentration of sodium acetate.

In the presence of sodium chromate as an inorganic inhibitor in NaOH solution, it is shown that (–ΔG) values are arranged in the following sequence:

  -ΔG in pH=13

at constant Temp.   *pure Al > 5083 > 7075 > 2024*

   and [CrO₄²⁻]

-ΔG in pH=9
at constant Temp.　　　　**pure Al > 7075 > 5083 > 2024**
and [CrO₄²⁻]

while various behaviour are observed in pH=11.

2- Values of ΔS are generally positive due to negativity of ΔG, this suggests a greater order in the solvated states of the metal ions as compared with the state of metal atoms in the crystal lattice of the corroding electrodes.

3- Generally, (ΔH) values are negative in pH=13 for pure Al and its alloys, which may be arranged in the following sequence:

　　　(-ΔH) value　　　*7075 > 5083 > pure Al > 2024*

While (ΔH) values are positive in pH=9 for pure Al and its alloys, as shown in the following sequence:

　　　(+ΔH) value　　　*7075 > 5083 > 2024 > pure Al*

But in pH=11, (ΔH) values are negative for pure Al and 5083 alloy, and positive for 2024 and 7075 alloy, as shown in the following sequence :

　　　(ΔH) value　　　*7075 > 2024 > pure Al > 5083*

In the presence of NaCl in solution, (ΔH) are negative for pure Al, 5083, and 7075 alloy while positive for 2024 alloy in pH=13 and 11. But in pH=9, (ΔH) values are positive for pure Al and its alloys.

In presence of CH_3COONa in solution, (ΔH) values are negative for pure Al, 5083, and 7075 alloy, while positive for 2024 alloy in PH=13 and 11. But in PH=9, there are various behaviour of (ΔH).

In the presence of Na_2CrO_4 , (ΔH) values are negative for pure Al and its alloys in pH=13, but (ΔH) values are negative for 2024 and 7075 alloy and take various behaviour for pure Al and 5083 alloy in pH=11.

While in pH=9, (ΔH) values are negative for pure Al, positive for 2024 alloy, and different for 5083 and 7075 alloys.

Generally, negative values of (ΔH) indicate a stronger bonding of the metal ions, resulting from electrode corrosion, with the species that are present in the corrosion medium as compared with the bonding of the metal atoms in the crystal lattice of the solid electrode.

4- The (E_a) values in NaOH solution in the absence of additives may be arranged in a sequence as:

In pH=13   (E_a)   7075 > 2024 > 5083 > pure Al

In pH=11   (E_a)   2024 > pure Al > 7075 > 5083

In pH=9   (E_a)   pure Al > 2024 $\simeq$ 7075 > 5083

The highest value of activation energy is found (127.97 kJ.mol^{-1}) for 7075 alloy in pH=13 in the presence of (1×10^{-3} mol.dm^{-3}) NaCl, while the lowest value of E_a is found 4.49 for pure Al in pH=9 + 1×10^{-2} mol.dm^{-3} NaCl).

5- The (log A) values (ranged between positive and negative values) may also be summarized in the following :

In NaOH solution in the absence of additives may be arranged in the following sequence :

In pH=13 (log A)   2024 $\simeq$ 7075 > 5083 > pure Al

In pH=11 (log A)   pure Al > 2024 > 5083 > 7075

In pH=9   (log A)   5083 > pure Al > 7075 $\simeq$ 2024

Generally, in the presence of additives, (log A) values may be arranged in a sequence as (except some cases) :

(log A) for pure Al and

its alloys in NaOH solution $NaCl > CH_3COONa > Na_2CrO_4$

in presence of additives

The highest value of **log A** was found (17.42) in pH=13 in presence of $(1 \times 10^{-3}$ mol.dm$^{-3})$ NaCl, while the lowest value of **log A** was found -6.58 (for 7075 alloy in pH=9 + 1×10^{-2} mol.dm^{-3} NaCl).

6- Generally, in the presence of additives, (ΔS^*) values may be arranged in a following sequence (except some cases):

(ΔS^*) for pure Al and

its alloys in NaOH solution $NaCl > CH_3COONa > Na_2CrO_4$

in presence of additives

7-2 Suggestions For Further Research

Several suggestions may be forwarded for further research on pure Al and its alloys may be summarized as:

1- The corrosion medium may be extended to other bases and acids in the absence and the presence of a number of organic and inorganic inhibitors.

2- The corrosion medium may also be subjected to thorough chemical analysis after corrosion tests to estimate the amounts of various metallic ions that can be formed through anodic dissolution of the working electrode. The working electrodes may also be examined carefully by scanning electron microscope and ESCA techniques subsequent to the various corrosion tests.

3- Other aluminium – base alloys may be covered in the corrosion experiments. These may be involve alloys containing metals other than Cu, Zn, Mg.

4- The experimental program may be repeated using different ions, i.e, F$^-$, SO$_4^{2-}$, and organic inhibitors to observe the effect of these ions and of the additives on the electrochemical behaviour of aluminium and its alloys.

5- The experimental program may be repeated using stirring on corrosion under laminar flow condition, and the work may be examining many of different stirring rates.

6- The dry – corrosion behaviour of aluminium and its alloys may be examined over a wider range of temperatures by exposing their clean surfaces, under high vacuum – conditions, to such gases as SO_2 and H_2O.

The results would enable a comparison to be made with results obtained in this or similar works of the wet – corrosion of the metal and of its alloys.

7- An attempt may be made to protect the metal and the alloys surfaces using paints, ceramics and other protective prior to the electrochemical investigations.

8- Polarization behaviour for metal and its alloys may be repeated using deaerated and oxygenated solutions.

References

1- H. P. Hoar, *J. Appl. Chem.*, vol.**11**, pp.121, **(1961)**.

2- R. F. Stegerwald, *Corrosion*, vol.**24**, pp.1, **(1968)**.

3- H. H. Uhlig, **"The Corrosion Handbook"**, (Wiley, New York, and Chapman and Hall, London), **(1948)**.

4- W. H. Ailor, **"Handbook on Corrosion Testing and Evaluation"**, (John Wiley and Sons), 8:171, **(1971)**.

5- C. Wagner and Traud, *Z. Electrochem.*, vol.**44**, pp.391, **(1938)**.

6- L. L. Shrier, **"Corrosion"**, **Corrosion Control**, (Newness- Butterworths, Boston), vol.**2**, pp.20:39, **(1978)**.

7- J. Z. Tafel, *Physik Chem.*, vol.**50**, pp.641, **(1904)**.

8- K. B. Oldham and F. Mansfeld, *Corrosion*, vol.**27**, pp.434, **(1971)**.

9- K. B. Oldham and F. Mansfeld, *Corrosion Sci.*, vol.**11**, pp.787, **(1971)**.

10- J. O'. M. Bockris and A. K. N. Reddy, **"Modren Electrochemistry"**, Phenam press, New York, vol.**2**, **(1970)**.

11- M. Stern and A. L. Grary, *J. Electrochem Soc.*, vol.**56**, pp.104, **(1957)**.

12- M. Stern and E. D. Weisert, *Proc. Am. Soc.* Test Master, vol.**59**, pp.1280, **(1959)**.

13- L. M. Callow, J. A. Ridiardson, and J. L. Dawson, *British Corrosion J.*, vol.**11**, pp.132, **(1976)**.

14- S. Barnartt, *Electrochim Acta*, vol.**10**, pp.1313, **(1970)**.

15- S. Barnarr, *Corrosion*, vol.**27**, pp.467, **(1971)**.

16- J. C. Reeve and G. Bech – Nielsen, *Corrosion Sci.*, vol.**13**, pp.351, **(1973)**.

17- K. B. Oldham and F. Mansfeld, *Corrosion Sci.*, vol.**14**, pp.813, **(1973)**.

18- M. Periassamy and P. R. Karishnaswamy, *J. Electronal Chem.*, vol.**61**, pp.349, **(1975)**.

19- J. Janko Wisk and R. Juchniewicz, *Corrosion Sci.*, vol.**20**, pp.841, **(1980)**.

20- I. Qamor and S. W. Hausion, *British Corrosion J.*, vol.**25**, pp.202, **(1990)**.

21- J. R. Davis, **"Metal Handbook"**, second edition **(1998)**, pp. 417.

22- R. V. Blewett and E. W. Skerry, *Metallurgia*, vol.71, No.73, **(1965)**.

23- H. P. Godard, **"The Corrosion of Light Metals"**, New York, J. Wiley and Sons, **(1967)**.

24- W. K. Johnson and T. E. Wright, *Aluminium*, vol.43, pp.490, **(1967)**.

25- J. R. Galvele and DE DE Micheli, *Corrosion Sci.*, vol.10, pp.795, **(1970)**.

26- I. L. Muller and J. R. Galvele, *Corrosion Sci.*, vol.17, pp.179, **(1977a)**.

27- L. Peralod Bicelli, C. Romagnoni, and D. Sinigaglia, *Corrosion Sci.*, vol.19, pp.553, **(1979)**.

28- P. L. Bonora, G. P. Ponzano, and V. Lorenzelli, *Br. Corrosion J.* (Quarterly), vol.2, pp.108, **(1974-I)**.

29- I. L. Muller and J. R. Galvele, *Corrosion Sci.*, vol.17, pp.995, **(1977b)**.

30- H. Bohni and H. H. Uhlig, *J. Electrochem. Soc.*, vol.116, pp.906, **(1969)**.

31- Norvald Nilsen and Einar Bradal, *Corrosion Sci.*, vol.17, pp.635, **(1977)**.

32- R. L. Horst and G. C. English, *Materials Performance*, vol.16, No.23, **(1977)**.

33- S. H. Sanad, A. A. Ismael, K. M. El-Sobki, and L. A. Shalapy, *Corrosion Prev. and Cont.*, vol.29, No.21, **(1982)**.

34- H. Herbert Uhlig, Ph.D, **"The Corrosion Handbook"**, pp. 39, **(1948)**.

35- L. L. Sherir, **"Corrosion"**, **Metal/Environment Reactions**, second ed. vol.1, pp. 4:12, **(1976)**.

36- M. DE Stella, *Corrosion Sci.*, vol.18, pp. 605-616, **(1978)**.

37- K. Nisancioglu and H. Holtan, *Corrosion Sci.*, vol.18, pp.835-849, **(1978)**.

38- K. Nisancioglu and H. Holtan, *Corrosion Sci.*, vol.18, pp.1011-1023, **(1978)**.

39- Zaki Ahned, *Anti- Corrosion*, pp.4-7, June **(1981)**.

40- Zaki Ahned, *Anti- Corrosion*, pp.4-10, July **(1981)**.

41- R. C. Salvarezza, M. F. L. Mele, and H. A. Videla, *Br. Corrosion J.*, vol.16, No.3, pp.162-168, **(1981)**.

42- R. C. Mc. Cune, R. L. Shilts, and S. M. Ferguson, *Corrosion Sci.*, vol.22, No.11, pp.1049-1065, **(1982)**.

43- P. Mauret and P. Lacaze, *Corrosion Sci.*, vol.**22**, No.**4**, pp.321-329, **(1982)**.

44- S. H. Sanad, A. A. Ismail, K. M. El-Sobki and L. A. Shalaby, *Corrosion Prev. and Cont.*, vol.**29**, No.**21**, pp.21-23, June **(1982)**.

45- K. Schwabe, S. Herrmann and F. Berthold, *Corrosion Sci.*, vol.**23**, No.**3**, pp.261-269, **(1983)**.

46- Boguslaw Mazurkiewz, *Corrosion Sci.*, vol.**23**, No.**7**, pp.687-690, **(1983)**.

47- Boguslaw Mazurkiewicz and Antonio Piotrowski, *Corrosion Sci.*, vol.**23**, No.**7**, pp.697-707, **(1983)**.

48- A. Csanady, T. Turmezey, I. Imre-Baa, A. Griger, D. Marton, L. Foder and L. Vitalis, *Corrosion Sci.*, vol.**24**, No.**3**, pp.237-248, **(1984)**.

49- F. Hunkeler and H. Bohni, *Corrosion – nace*, vol.**40**, No.**10**, pp.534-540, October **(1984)**.

50- W. Michael, Moore, Chia-tien Chen, and A. George Shirn, *Corrosion – nace*, vol.**40**, No.**12**, pp.644-649, December **(1984)**.

51- B. A. Abo-El-Nabey, N. Khalil, and E. Khamis, *Corrosion Sci.*, vol.**25**, No.**4**, pp.225-232, **(1985)**.

52- A. I. Onuchukwu and F. K. Oppong – Boachie, *Corrosion Sci.*, vol.**26**, No. **11**, pp.919-926, **(1986)**.

53- zaki Ahmed, *Anti- corrosion*, pp.4-11, November **(1986)**.

54- P. L. Cabot, J. A. Garrido, E. Perez and J. Virgili, *Corrosion Sci.*, vol.**26**, No.**5**, pp.357-369, **(1986)**.

55- W. C. Moshier, G. D. Davis and J. S. Ahearn, *Corrosion Sci.*, vol.**27**, No.**8**, pp.785-801, **(1987)**.

56- E. Lunarska, and Z. Szklarska – Smialowska, *Corrosion – nace*, vol.**43**, No.**6**, pp.353-359, June **(1987)**.

57- E. Lunarska, and Z. Szklarska – Smialowska, *Corrosion – nace*, vol.**43**, No.**7**, pp.414-424, July **(1987)**.

58- A. A. Mazhar, F. El-Taib Heakal, and A. S. Mogoda, *Corrosion – nace*, vol.**44**, No.**6**, pp.354-359, june **(1988)**.

59- F. Qvari, L. Tomesanyi, and T. Turmezey, *Electrochimica Acta*, vol.33, No.3, pp.323-326, **(1988)**.

60- V. Surganov, C. Jansson, J. G. Nielesn, and P. Morgen, *Electrochimica Acta*, vol.33, No.4, pp.517-519, **(1988)**.

61- Hung-Pyo Kim, Rak-Hyun Song, and Su-Il Pyun, *Br. Corrosion J.*, vol.23, No.4, **(1988)**.

62- B. Tsujio and T. Oki, *Corrosion Sci.*, vol.44, No.12, pp.900-905, December **(1988)**.

63- T. C. Tan and D. T. Chin, *Corrosion*, vol.45, No.12, pp.984-989, December **(1989)**.

64- L. Tomesanyi, K. Varga, I. Bartik, G. Horanyi and E. Maleczki, *Electrochimica Acta*, vol.34, No.6, pp.855-859, **(1989)**.

65- S. Furuya and N. Soga, *Corrosion Engineering*, vol.39, No.2, pp.79-87, **(1990)**.

66- M. elboujdaini and E. Ghali, *Corrosion Sci.*, vol.30, No.8/9, pp.855-867, **(1990)**.

67- T. Xue, W. C. Cooper, R. Pascual, S. Saimoto, *J. Appl. Electrochemistre*, vol.21, pp.238-246, **(1991)**.

68- F. Holzer, S. Muller, j. Desilvestro, O. Haas, *J. Appl. Electro.*, vol.23, pp.125-134, **(1993)**.

69- S. B. Saidman, S. G. Garcia and J. B. Bessone, *J. Appl. Electro.*, vol.25, No.3, pp.252-258, March **(1995)**.

70- J. William, D. Jeremy, L. Richard, and S. Gerald, *J. Electro. Soc.*, vol.149, No.5, pp.179-185, **(2002)**.

71- M. N. Desai, B. C. Thakar, P. M. Chhaya and M. H. Gandh, *Corrosion Sci.*, vol.16, pp.9-24, **(1976)**.

72- W. J. Rudd and J. C. Scully, *Corrosion Sci.*, vol.20, pp.611-631, **(1980)**.

73- J. D. Talati, G. A. Patal and b.P. Patal, *Br. Corrosion J.*, vol.15, No.2, pp.85-88, **(1980)**.

74- B. M. Abo-El-Khair and B. G. Ateya, *Corrosion Prev. and Cont.*, pp;7-9, August **(1981)**.

75- D. D. N. Singh and C. V. Agarwal, *Corrosion Prev. and Cont.*, pp;11-15, August **(1982)**.

76- J. D. Talati and D. K. Gandhi, *Corrosion Sci.*, vol.**23**, No.**12**, pp.1315-13332, **(1983)**.

77- Abo El-Khair and B. Mostafa, *Corrosion Prev. and Cont.*, pp;15-17, February **(1983)**.

78- P. N. S. Yadav, R. S. Chaudhary and C. V. Agarwal, *Corrosion Prev. and Cont.*, pp;9-13, October **(1983)**.

79- C. Chakrabarty, M. M. Singh and C. V. Agarwal, **Br. Corrosion J.**, vol.**18**, No.**2**, pp;107-110, **(1983)**.

80- J. D. Talati, G. A. Patel, and D. K. Gandhi, *Corrosion – nace*, vol.**40**, No.**2**, pp.88-92, Feberuary **(1984)**.

81- Abo. El-khair and B. Mostafa, *Corrosion Prev. and Cont.*, pp.17-19, October **(1984)**.

82- Wafia El Sayed, *Corrosion Prev. and Cont.* pp.16-19, December **(1984)**.

83- H. A. Dessoki, A. A. Abdel Fattah and S. M. Abd El Haleem, *Corrosion Prev. and Cont.*, pp;18-22, August **(1984)**.

84- A. S. Fouda, M. N. Moussa, F. I. Taha and A. I. Elneanaa, *Corrosion Sci.*, vol.**26**, No.**9**, pp.719-726, **(1986)**.

85- I. Ahmed, S. N. Basahel and R. M. Khalil, *Anti – corrosion*, pp.4-8, August **(1988)**.

86- D. R. Arnott, B. R. W. Hinton, and N. E. Ryan, *Corrosion*, pp.12-18, January **(1989)**.

87- M. M. El-Tagouri and M. R. Mostafa, *Anti – corrosion*, pp.10-14, September **(1989)**.

88- S. Sankarapapavinasam, F. Pushpanaden, M. F. Ahmed, *J. Appl. Electro.*, vol.**21**, pp.625-631, **(1991)**.

89- J. Radosevic, M. Kliskic, and A. R. Despic, *J. Appl. Electro.*, vol.22, pp.649-656, **(1992)**.

90- M. Metikos – Hukovic, R. Babic, Z. Grubac, and S. Brinic, *J. Appl. Electro.*, vol.**24**, pp.325-331, **(1994)**.

91- S. Frankel and L. Richard, *The Electro. Soc.*, Winter **(2001)**.

92- D. Ornek, A. Jayaraman, B. C. Syrett, C. H. Hsu, F. B. Mansfeld, T. K. Wood, *Appl. Microbiol. Biotechnol.*, vol.**58**, pp.651-657, **(2002)**.

93- M. Luis Enrique, W. Joao Fabio, V. Idalina, G. Hercilio, *J. Braz. Chem. Soc.*, vol.**14**, No.**4**, July/Aug **(2003)**.

94- J. A. Van Frawnhofer and G. H. Banks, **"Potentiostat And Its Applications"** (Butter Worth, London), **(1972)**.

95- J. O. M. Bockris and A. K. N. Reddy, **"Modern Electrochemistry"** , (Phenam Press, New York), vol.**2**, pp.1315, **(1970)**.

96- **Instruction Manual Potentiostat**, SOLEA. TACUSSEL, France, **(1980)**.

97- **Instruction Manual Potentiometric Laboratory Recorder**, SOLEA. TACUSSEL, **(1973)**.

98- **Instruction Manual DC Digital Millivoltmeter - Volttmeter**, SOLEA. TACUSSEL, **(1980)**.

99- **American Society for Testing and Material Annual Book of ASTM Standard**, (1978-G5).

100- J. M. Saleh and Y. K. Al-Haidari, *Bull. Chem. Soc. Jpn.*, vol.**62**, pp.1237, **(1989)**.

101- T. Donald Sawyer, **"Experimental Electrochemistry For Chemists"**, (Wiley, New York), pp.45, **(1974)**.

102- T. Lymon (ed), **"Metals Handbook"**, (ASM, Ohio), pp.159, **(1958)**.

103- L. M. Al-Shamma, I. M. Saleh and N. A. Hikmat, *Corrosion Sci.* vol.**27**, pp.22, **(1987)**.

104- N. A. Hikmat, Ph. D. Thesis, College of Science, University of Baghdad, January, **(2002)**.

105- I. D. Sulliman, Ph. D. Thesis, College of Education Ibn Al-Haitham, University of Baghdad, November, **(2003)**.

106- N. D. Green, **"Experimental Electrode Kinetics"**, (Troy, New York), pp.64, **(1965)**.

107- J. W. D. France, *Mat. Res.*, vol.9, No.21, **(1969)**.

108- L. M. Al-Samma, M. Sc. Thesis, college of Science, University of Baghdad, June, **(1985)**.

109- G. B. Roger, **"Determination of pH"**, (Wiley, New York), pp.307, **(1972)**.

110- **American Society for Testing and Material Annual Book of ASTM Standard**, part 10, **(1980)**.

111- T. P. Hoar, D. C. Mears, and G. P. Rothwell, *Corrosion Sci.*, vol.5, pp.279, **(1965)**.

112- C. R. Schmitt, *Rev. Coat. Corrosion*, vol.4, No.1, pp.97, **(1995)**.

113- R. H. Hart, *J. Electrochem. Soc.*, vol.30, pp.57, **(1957)**.

114- M. Heine, D. S. Keir, and M. J. Pryor, *J. Electrochem. Soc.*, vol.24, No.1, pp.112, **(1965)**.

115- W. E. Ruther, and J. E. Draley, *Corrosion*, vol.12, pp.31, **(1956)**.

116- A. C. Fraker, and A. W. T. Ruff, *Corrosion – nace*, vol.27, No.4, pp.151, **(1971)**.

117- D. F. Maclehnan, *Corrosion*, vol.17, pp.105, **(1961)**.

118- V. H. Trouther, *Corrosion*, vol.15, pp.9t, **(1951)**

119- Becerra Alcibides, and Darby Ron, *Corrosion – nace*, vol.30, No.5, pp.153, **(1975)**.

120- S. Maitra and G. C. English, *Metallurgical Transaction A*, vol.12A, pp.535, **(1981)**.

121- G. Fontana and D. Green, **"Corrosion Engineering"**, Mc Graw – Hill, New York, **(1978)**.

122- R. Pierre Roberge, **"Handbook of Corrosion Engineering"**, Mc Graw – Hill, New York, **(1999)**.

123- H. Y. Chen, *Journal of Power Sources*, vol.88, No.78, **(2000)**.

124- E. Rocca, *J. Electronal. Chem.*, vol.543, No.153, (2003).

125- H. H. Uhlig, **"Corrosion and Corrosion Control"**, Wiley, New York, (2000).

126- L. M. Al-Shama`a, Ph.D., <u>Thesis</u>, College of Science, University of Baghdad, April, (1999).

127- J. O. M. Bockeris and A. K. Reddy, **"Modern Electrochemistry"**, Press, New York, pp.176, (1970).

128- E. Heitz and W. Schwenk, *Br. Corrosion J.*, vol.11, No.2, (1976).

129- Trung Hung Nguyen and R. T. Foley , *J. Electrochem. Soc.*, vol.126, pp.1855, (1979).

130- H. J. Engell, *Electrochim. Acta*, vol.22, pp.987, (1977).

131- J. kruger and C. Mc Bee, **"Localized Corrosion"**, pp.252, nace – 3, (1974).

132- T. P. Hoar, *Corrosion Sci.*, vol.7, pp.341, (1967).

133- Augustynski, **"Passivity of Metals"**, pp.989, (1978).

134- J. Painot and J. Augustynski, *Electrochim. Acta.*, vol.20, pp.747, (1975).

135- J. Painot and J. Augustynski, *J. Electrochem. Soc.*, vol. 123, pp.84, (1976).

136- M. Koudelkova and J. Augustynski, *J. Electrochem. Soc.*, vol. 126, pp.1659, (1979).

137- M. Koudelkova, J. Augustynski, and H. Berthon, *J. Electrochem. Soc.*, vol.124, pp.1165, (1977).

138- A. F. Beck, M. A. Heine, D. S. Keir, D. Van Rooyen, and M. J. Pryor, *Corrosion Sci.*, vol.2, pp.123, (1962).

139- M. A. Heine, D. S. Keir, and M. J. Pryor, *J. Electrochem. Soc.*, vol.112, pp.24, (1965).

140- Z. Szklarska – Smialowska, *Corrosion*, vol.27, pp.223, (1971).

141- JA. M. Kolotyrkin, *Corrosion*, vol.19, pp.261t, (1963).

142- JA. M. Kolotyrkin, *J. Electrochem. Soc.*, vol.108, pp.209, (1961).

143- G. M. Schmidt and N. Hackerman, *J. Electrochem. Soc.*, vol.108, pp.741, (1961).

144- H. H. Uhlig, *J. Electrochem Soc.*, vol.97, pp.215, (1950).

145- F. P. Robinson, *Corrosion Technol.*, vol.7, pp.266, **(1960)**.

146- I. L. Rozenfeld and I. K. Moshokov, *Corrosion*, vol.20, pp.115, **(1964)**.

147- H. J. Vetter and H. H. Strehblow, **"Localized Corrosion"** nace-3, pp.240, **(1974)**.

148- J. A. Richardson and G. C. Wood, *J. Electrochem. Soc.*, vol.120, pp.193, **(1973)**.

149- M. F. Abd Rabbo, G. C. Wood , J. A. Richardson, and C. K. Jackson, *Corrosion Sci.*, vol.14, pp.645, **(1974)**.

150- M. F. Abd Rabbo, G. C. Wood , J. A. Richardson, and C. K. Jackson, *Corrosion Sci.*, vol.16, pp.677, **(1976a)**.

151- M. Janki – Ceachor, G. C. Wood, and G. E. Thompson, *Br. Corrosion J.*, vol.15, pp.154, **(1980)**.

152- G. Wranglen, **"Introduction to Corrosion and Protection of Metals"**, Butter and Tanner, London, **(1972)**.

153- M. Janik – Czachor, A. Czummir, and Szklarska – Smialowska, *Corrosion Sci.*, vol.15, pp.775, **(1975)**.

154- T. P. Hoar and W. R. Jacob, *Nature*, **vol.216**, pp.1299, **(1967)**.

155- K. F Lorking and J. E. O. Mayna, *J. App. Chem.*, vol.20, pp.611, **(1961)**.

156- G. C. Natha, **"Corrosion Inhibitor"**, Nace, Texai, **(1973)**.

157- K. R. Trethewey and J. Chamberlain, **"Corrosion for Science and Engineering"**, 2nd ed. , Addison Wesley Longman Ltd, **(1996)**.

158- I. L. Rozenfeld, **"Corrosion Inhibitor"**, Mc Graw Hill, New York, **(1981)**.

159- H. Fischer, **"Definition and Modes of Inhibition of Electrochemical Electrode Reaction in Proceeding of 3rd European Symposia on Corrosion Inhibitors"**, Ferrara, Italy, pp.15-21, **(1970)**.

160- S. M. Mayanna and T. H. Y. Setty, *Corrosion Sci.*, vol.15, pp.**627**, **(1975)**.

161- M. E. Straumanis and N. Brakss, *Metallen*, vol.3, pp.41, **(1949)**.

162- I. N. Putilova, V. P. Barannik, S. A. Balezin, **"Metallic Corrosion Inhibitors"**, Pergamon Press, Oxford, pp.121, **(1960)**.

163- L. I. Antropov, **"Proc. 1ˢᵗ. Int. Congr. Metallic Corrosion"**, London, pp.148, **(1962)**.

164- A. M. Farhan, Ph. D. <u>Thesis</u>, College of Science, University of Baghdad, June, **(2000)**.

165- C. Edeleanu and U. R. Evans, ***Trans. Faraday Soc.***, vol.47, pp.1121, **(1951)**.

166- M. A. Heine and M. J. Pryor, ***J. Electrochem. Soc.***, vol.114, pp.1001, **(1969)**.

167- K. R. Tretherwey and J. Chamberlain, ***Corrosion for Science and Engineering***, 2ⁿᵈ ed.,(Addision Wesley Longman Ltd.), **(1996)**.

168- S. A. H. Heswa, Ph. D. <u>Thesis</u>, College of Science, Sadam University, July, **(1996)**.

169- Y. K. Al-Haydari, Ph. D. <u>Thesis</u>, College of Science, University of Baghdad, January, **(1998)**.

170- L. M. Al- Shamma, J. M. Saleh, and N. A. Hikmat, ***Corrosion Sci.***, vol.27, pp.22, **(1987)**.

171- J. M. Saleh, K. A. Saleh, and N. A. Hikmat, ***Iraqi J. Sci.***, vol.36, pp.803, **(1995)**.

172- G. Cbond, **"Catalysis by Metals"** (Academic Press, New Yourk) pp. 70-126 and 140, **(1962)**.

173- Y. K. Al-Haydari, J. M. Saleh and M. H. Matloob, ***J. Phys. Chem.***, vol.89, pp.3286, **(1985)**.

174- S. A. Isa and J. M. Saleh, ***J. Phys. Chem.***, vol.76, pp.2530, **(1972)**.

175- E. Cremer, **"Advance in Catalysis"**, (Academic press, New York), **(1955)**.

Appendix

Composition limits for wrought aluminium and aluminium alloys which are shown in this thesis[21].

Aluminium And its alloys	Si	Fe	Cu	Mn	Mg	Cr	Ni	Zn	Ga	V
1060	0.25	0.35	0.05	0.03	0.03	-	-	0.05	-	0.05
1100	0.95 Si+Fe		0.05-0.2	0.05	-	-	-	0.10	-	-
1199	0.006	0.006	0.006	0.002	0.006	-	-	0.006	0.005	0.005
2017	0.2-0.8	0.7	3.5-0.2	0.4-1	0.4-0.8	0.1	-	0.25	-	-
2024	0.5	0.5	3.8-4.9	0.3-0.9	1.2-1.8	0.1	-	0.25	-	-
3003	0.6	0.7	0.05-0.2	1-1.5	-	-	-	0.1	-	-
5052	0.25	0.4	0.1	0.1	2.2-2.8	-	0.15-0.35	0.1	-	-
5082	0.2	0.35	0.15	0.15	4-5	0.15	-	0.25	-	-
5083	0.4	0.4	0.1	0.4-1	4-4.9	0.05-0.25	-	0.25	-	-
5154	0.25	0.4	0.1	0.1	3.1-3.9	0.15-0.35	-	0.2	-	-
5454	0.25	0.4	0.1	0.5-1	2.4-3	0.05-0.2	-	0.25	-	-
5456	0.25	0.4	0.1	0.5-1	4.7-5.5	0.05-0.2	-	0.25	-	-
5457	0.08	0.1	0.2	0.15-0.45	0.8-1.2	-	-	0.05	-	0.05
6061	0.4-0.8	0.7	0.15-0.4	0.15	0.8-1.2	0.04-0.35	-	0.25	-	-
7016	0.1	0.12	0.45-0.1	0.03	0.8-1.4	-	-	4-5	-	0.05
7021	0.25	0.4	0.25	0.1	1.2-1.8	0.05	-	5-6	-	-
7029	0.1	0.12	0.5-0.9	0.03	1.3-2	-	-	4.2-5.2	-	0.05
7039	0.3	0.4	0.1	0.1-0.4	2.3-3.3	0.15-0.25	-	3.5-4.5	-	-
7075	0.4	0.5	1.2-2	0.3	2.1-2.9	0.18-0.28	-	5.1-6.1	-	-

Lightning Source UK Ltd.
Milton Keynes UK
UKHW010645080421
381649UK00001B/31